A First
Course in
Functional
Analysis

A First Course in Functional Analysis

MARTIN DAVIS
PROFESSOR EMERITUS
COURANT INSTITUTE OF
MATHEMATICAL SCIENCES
NEW YORK UNIVERSITY

DOVER PUBLICATIONS, INC.
MINEOLA, NEW YORK

Bibliographical Note

This Dover edition, first published in 2013, is an unabridged republication of the work originally published by Gordon and Breach, New York, in 1966.

Library of Congress Cataloging-in-Publication Data

Davis, Martin, 1928–
    A first course in functional analysis / Martin Davis, professor emeritus, Courant Institute of Mathematical Sciences, New York University.
        p. cm.
    Originally published: New York : Gordon and Breach, 1966, in the series Notes on mathematics and its applications.
    Includes index.
    ISBN-13: 978-0-486-49983-3
    ISBN-10: 0-486-49983-9
    1. Functional analysis. I. Title.

QA320.D32 2013
515'.7—dc23

2012041596

Manufactured in the United States by Courier Corporation
49983901
www.doverpublications.com

To Harold and Nathan

Preface

The undergraduate mathematics major nowadays encounters modern mathematics as a collection of more or less unconnected subjects. The aim of this little book is to demonstrate the essential unity of twentieth-century mathematics without assuming more mathematical knowledge or maturity than can reasonably be presumed of seniors or beginning graduate students.

The contents may be described as an exposition, starting from scratch, of Gelfand's proof via maximal ideals of Wiener's famous result that when an absolutely convergent trigonometric series has a non-vanishing sum, the reciprocal of the sum can likewise be expanded into an absolutely convergent trigonometric series. En route one is led to prove Zorn's lemma and the Hahn-Banach extension theorem and to find the geometric series representation of inverces in an Abelian-Banach algebra. And each of these leads to brief detours: to a discussion of the functional equation $f(x+y) = f(x) + f(y)$, of the existence of Green's functions, and of existence and uniqueness theorems for certain linear integral and differential equations.

The book is almost entirely self-contained. The reader should have had some experience with $\varepsilon - \delta$. A prerequisite (or corequisite) would be complex analysis through Cauchy's integral formula, unless the instructor is willing to take the extra time needed to develop this material. Also it would be advisable for the student to have seen the quotient group or quotient ring construction before encountering it here.

I taught a three-credit-hour course covering the present material during the Spring 1959 semester at the Hartford Graduate Center of Rensselaer Polytechnic Institute to beginning graduate students with quite modest preparation. No suitable text being available, dittoed lecture notes were prepared from the students' own notes, the students, in turn, each being responsible for a specific portion of the course. The present book consists of these very notes, lightly edited. The problems were taken from those assigned as homework as well as from the final examinations.

I am grateful to Jack Schwartz for suggesting that these notes be included in the series under his editorship, and I will feel most pleased if this book leads others to enjoy Gelfand's beautiful proof.

MARTIN DAVIS
New York City
April 1966

Contents

CHAPTER I

Set Theoretic Preliminaries

I. Sets and Members

In this section the terms class, set, collection, totality and family will be used synonomously. The symbol, $\in$ (epsilon), will be used to denote membership in a class. The symbol, $\notin$, will denote non-membership in a class, i.e.,

$$x \in C \text{ means } x \text{ is a member of set } C$$

$$x \notin C \text{ means } x \text{ is not a member of set } C$$

Example: If C is the class of all even numbers, then 2 belongs to C, $(2 \in C)$, 4 belongs to C, $(4 \in C)$, and 3 does not belong to C, $(3 \notin C)$.

If a set is finite, then it can be described by listing its members. For example:

Let C be the set $\{1, 3, 5\}$ then $1 \in C$, $3 \in C$, $5 \in C$ and all other elements are not members of the set C. The description of an infinite set, however, is not so simple. An infinite set can not be described by simply listing its elements. Hence, we must have recourse to defining the set by a characteristic property. The set can then be described as the set of all elements which possess the property in question. If the property is, say, $P(x)$, then the set will be written $\{x \mid P(x)\}$, thus to write $C = \{x \mid P(x)\}$, is to say that C is the set of all elements x, such that, $P(x)$ is true. For example,

$$\{a, b\} = \{x \mid x = a \text{ or } x = b\}$$

A set may have only one element: $\{a\} = \{x \mid x = a\}$.

Definition I.I. The *union* of set A with set B, $A \cup B$ is the set of all those elements which belong either to set A or to set B or to both. Symbolically,

$$A \cup B = \{x \mid x \in A \text{ or } x \in B\}$$

1

Definition 1.2. The *intersection* of two sets, $A \cap B$, is defined to be the set of all those elements which are common to both set A and set B. Symbolically,

$$A \cap B = \{x \mid x \in A \text{ and } x \in B\}$$

Definition 1.3. The *difference* between two sets, denoted by $A - B$ is defined to be the set

$$A - B = \{x \mid x \in A \text{ and } x \notin B\}$$

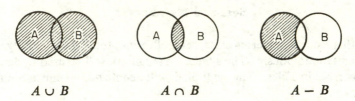

$$A \cup B \qquad\qquad A \cap B \qquad\qquad A - B$$

We shall use the symbols:

$\Rightarrow$ to mean "implies"

$\Leftrightarrow$ to mean "if and only if" and

$\exists$ to mean "there is a"

Actually a certain amount of care is necessary in using the operation: $\{x \mid \ldots \ldots\}$. E.g., let $P(x)$ be the property: $x \notin x$. (Thus, we could plausibly claim as an x for which $P(x)$ is true, say, the class of even numbers, whereas the class of, say, non-automobiles, or the class of entities definable in less than 100 English words, could be claimed as x's making $P(x)$ false.) Suppose we can form:

$$A = \{x \mid x \notin x\}$$

i.e.
$$x \in A \Leftrightarrow x \notin x$$

Applying this to $x = A$, yields

$$A \in A \Leftrightarrow A \notin A$$

a contradiction. This is Russell's paradox.

In an axiomatic treatment of set theory (cf. Kelley, *General Topology*, Appendix) suitable restrictions have to be placed on the operation:

$\{x \mid \ldots \quad \ldots\}$ in order to prevent the appearance of the Russell paradox or similar paradoxes. In these lectures we shall use this operation whenever necessary. However, all of our uses could be justified in axiomatic set theory.

The symbol $=$ of equality always will mean absolute identity. In particular, for sets A, B, the assertion $A = B$ means that the sets A and B have the same members.

Definition 1.4. The set A is a *subset* of a set B, written $A \subset B$, if each element of A is also an element of B, that is if $x \in A \Rightarrow x \in B$.

From this definition it follows that $A \subset A$.

Definition 1.5. The *empty set*, represented by the symbol $\emptyset$, is the set which contains no elements.

For example:

$$\emptyset = \{x \mid x \neq x\}$$

$$\emptyset = \{x \mid x = 0 \text{ and } x = 1\}$$

The empty set is a subset of any set, i.e. $\emptyset \subset A$.

Definition 1.6. 2^A is the class of all sets B, such that B is a subset of A, i.e.

$$2^A = \{B \mid B \subset A\}$$

For example, consider the set A, where

$$A = \{L, M, N\}$$

Then the elements of 2^A are the empty set $\emptyset$, the sets containing only one element, $\{L\}$, $\{M\}$, $\{N\}$, the sets containing two elements, $\{L, N\}$, $\{L, M\}$, $\{M, N\}$, and finally the single set containing all three elements $\{L, M, N\}$. Note that A contains 3 elements and that 2^A contains $8 = 2^3$ elements.

Definition 1.7. The *ordered pair* of two elements is defined as $(a, b) = \{ \{a\}, \{a, b\} \}$.

It can easily be shown that $(a, b) = (a', b') \Rightarrow a = a'$ and $b = b'$. (Cf. Problem 1.) It is this which is the crucial property of the ordered pair. Any other construct with this property could be used instead.

Note that (a, b) is quite different from $\{a, b\}$. For, $\{a, b\}$ is always equal to $\{b, a\}$.

Definition 1.8. The *Cartesian product* of sets A and B, written, $A \times B$, is the set of all ordered pairs (a, b), such that, a belongs to the set A, and b belongs to the set B, i.e.,

$$A \times B = \{(a,b) \mid a \in A \text{ and } b \in B\}$$

For example, if:

$$A = \{L, M, N\}$$

and

$$B = \{P, Q\}$$

then the Cartesian product is the set

$$\begin{Bmatrix} (L, P), & (L, Q) \\ (M, P), & (M, Q) \\ (N, P), & (N, Q) \end{Bmatrix} = A \times B$$

Note that there are 6 ($= 3 \times 2$) elements in the set.

2. Relations

Definition 2.1. A *relation between sets A and B* is a set C such that $C \subset A \times B$.

Examples of such a C are $\{(L, P), (L, Q)\}$, $\{(M, P), (M, Q)\}$, and $\{(N, P), (N, Q)\}$ as taken from the above example.

Definition 2.2. A *relation on A* is a relation between A and A.

Definition 2.3. A *mapping, transformation* or *function* α *from A into B* is a relation between A and B, such that, for each $x \in A$, there is *exactly one* $y \in B$ such that $(x, y) \in \alpha$.

We write: $\alpha(x) = y$ to mean $(x, y) \in \alpha$.

For example, let A be a set of positive integers and C be the set of ordered pairs (a, b) such that $a-b$ is even, i.e. $C = \{(a, b) \mid a-b$ is even$\}$. Then,

$$(1, 3) \ \in C$$
$$(2, 4) \ \in C$$
$$(1, 6) \ \notin C$$

Specifically, $(a, b) \in C \Leftrightarrow a \equiv b \bmod 2$. C is a relation on A.

Example: Let A be the set of human beings, and let α be the set of pairs (x, y) where $x \in A$, $y \in A$ and y is the father of x. Then, α is a relation on A and is also a function from A into A. However (cf. Def. 2.4 below), it is not a function from A onto A.

Definition 2.4. A function α is called a *mapping from A onto B*, if for each $y \in B$ there is *at least* one $x \in A$ such that $(x, y) \in \alpha$.

Definition 2.5. A function α is called a *one-one mapping from A onto B*, if for each $y \in B$, there is *exactly one* $x \in A$ such that

$$(x, y) \in \alpha$$

Example: Let R be the set of real numbers, and let P be the set of non-negative real numbers, let

$$\alpha = \{(x, y) \mid x \in R \text{ and } y = x^2\}$$

$$\beta = \{(x, y) \mid x \in P \text{ and } y = x^2\}$$

Then, α is a function from R into R. It is also a function from R into P and in fact is a function from R onto P. However α is not one-one. β is a one-one mapping from P onto P.

We should remark that, if α and β are functions, then $\alpha \subset \beta$ turns out to mean simply that

$$\alpha(x) = y \Rightarrow \beta(x) = y$$

In this case β is called an *extension* of α.

3. Equivalence Relations

As already noted, a relation on a set A is a subset of $A \times A$, so that a relation consists of ordered pairs of elements of A. The fact that an element $a \in A$ bears the relation R to $b \in A$ may be expressed in the form $(a, b) \in R$, or, as is more usually written, aRb.

Example: Let A be the set of all integers. Consider "$<$".

$$< \; = \{(a, b) \mid a \text{ less than } b\}$$

$$(2, 6) \in \; <, \text{ more usually written } 2 < 6$$

There are many kinds of relations, having many different properties. Three possible properties of relations are of special importance. A relation R on a set A is called:

> *reflexive* if xRx for all $x \in A$
>
> *symmetric* if $xRy \Rightarrow yRx$
>
> *transitive* if xRy and $yRz \Rightarrow xRz$

A relation R on a set A is called an *equivalence relation* on A if it is reflexive, symmetric, and transitive. Such a relation can always be replaced by the equality relation between suitable sets. If R is an equivalence relation on A, we define

$$[x] = \{y \in A \mid xRy\}$$

Theorem 3.1.

Let R be an equivalence relation on a set A. Then:

(1) $xRy \Leftrightarrow [x] = [y]$.

(2) $x \in [x]$.

(3) $[x] \cap [y] \neq \emptyset \Rightarrow [x] = [y]$.

(That is, the equivalence classes $[x]$ divide the set A in a manner such that:

(1) two elements bear the relation to each other if and only if they are in the same equivalence class;

(2) each element is in one equivalence class;

(3) the equivalence classes do not overlap.)

Proof: (2) $[x] = \{w \in A \mid xRw\}$

xRx (R is reflexive)

$\therefore \ x \in [x]$.

(3) Let $z_0 \in [x] \cap [y]$ and $t \in [x]$

$xRt, \ xRz_0, \ yRz_0$

$z_0 Rx$ (R is symmetric)

$z_0 Rt$ (R is transitive)

yRt (R is transitive)

$t \in [y]$

$[x] \subset [y]$.

The same argument, with x and y interchanged, may be used to show that $[y] \subset [x]$.

$$\therefore \ [x] = [y]$$

(1) $\Rightarrow$ $\Leftarrow$

    Let xRy Let $[x] = [y]$

    $y \in [x]$ $y \in [y]$ by (2)

    $y \in [y]$ by (2) $y \in [x]$ $([x] = [y])$

    $y \in [x] \cap [y]$ xRy

    $\therefore \ [x] = [y]$ by (3)

4. The Principle of Choice

This topic is controversial: the Principle of Choice will be proved using the Multiplicative Principle, which some people say is obvious and which others say is false. We shall use these principles freely throughout these lectures.

Convention: By a *family* we shall mean a collection of sets. Many sets are collections of sets, to be sure, but we shall use the term family when we wish to emphasize that its members are sets.

Definition 4.1. The family $\mathscr{F}$ of sets is called a *family of disjoint sets* if: $A \in \mathscr{F}$ and $B \in \mathscr{F}$ and $A \neq B \Rightarrow A \cap B = \emptyset$.

Example: $\{1, 2, 3\} \cap \{2, 4, 6\} = \{2\}$ so these two sets are not disjoint.

Example:

$$\mathscr{F} = \{ \ \{0, 3, 6, 9, \ldots\}$$
$$\{1, 4, 7, 10, \ldots\}$$
$$\{2, 5, 8, 11, \ldots\} \ \}$$

These three sets are disjoint. They are, indeed, equivalence classes (the relation being congruence modulo 3), and equivalence classes are disjoint (cf. Theorem 3.1 (3)).

The Multiplicative Principle: Let $\mathscr{F}$ be a family of non-empty disjoint sets. Then there is a set M which has exactly one element in common with each set of $\mathscr{F}$.

People object to the Multiplicative Principle: sometimes no principle of selection from each set can be stated, because the elements in the set can not be characterized in a usable fashion. Some people feel that this inability to specify the individual elements of M invalidates the principle; others feel that it does not.

Example: Take $A = \{x \in R \mid 0 \leq x \leq 1\}$ and xRy means that $x - y$ is rational. R is clearly an equivalence relation.

Here we have sets whose elements can be unnamable or even unimagined. Yet the multiplicative principle applies for the family $\mathscr{F}$ of all equivalence classes.

Theorem 4.1. (Principle of Choice).

Let $\mathscr{F}$ be a family of non-empty sets. Then there is a function f such that for each $A \in \mathscr{F}$, we have $f(A) \in A$.

Proof: For each $A \in \mathscr{F}$, define $T_A = \{(A, x) \mid x \in A\}$. Define

$$\mathscr{G} = \{T_A \mid A \in \mathscr{F}\}$$

$\mathscr{G}$ is a family of disjoint sets, so the multiplicative principle applies. There is a set f which has exactly one element in common with each T_A. This f satisfies the requirements of the theorem, since $y = f(A)$ means $(A, y) \in f$.

5. Zorn's Lemma

We define

$$\bigcup_{A \in \mathscr{F}} A = \{x \mid \text{ for some } A, x \in A \text{ and } A \in \mathscr{F}\}$$

$$\bigcap_{A \in \mathscr{F}} A = \{x \mid \text{ for all } A, A \in \mathscr{F} \Rightarrow x \in A\}$$

Definition 5.1. A *chain*, $\mathscr{C}$, is a non-empty family of sets such that $A \in \mathscr{C}$ and $B \in \mathscr{C} \Rightarrow A \subset B$ or $B \subset A$.

Example: $\{\{0\}, \{0, 1\}, \{0, 1, 2\}, \{0, 1, 2, 3\}\}$ is a chain but $\{\{0\}, \{0, 1\}, \{0, 2\}\}$ is *not* a chain.

Note that a family which is *not* a chain may have subsets which are chains. In the second example above, $\{\,\{0\},\ \{0,1\}\,\}$ and $\{\,\{0\},\ \{0,2\}\,\}$ are chains yet subsets of a family which is not a chain.

Definition 5.2. $\mathscr{F}$ is called an *inductive family* of sets if whenever $\mathscr{C}$ is a subset of $\mathscr{F}$ and $\mathscr{C}$ is a chain then $\bigcup_{A \in \mathscr{C}} A$ is in $\mathscr{F}$.

Examples of non-inductive sets:

(1) $\{\,\{0\}, \{0,2\}, \{0,2,4\}, \ldots, \{0,2,4,\ldots\}, \{0,2,4,\ldots,1\},$
$\{0,2,4,\ldots,1,3,\ldots\}\,\}$;

(2) $\{\,\{0\}, \{0,1\}, \{0,1,2\}, \{0,1,2,3\} \ldots\}$

Note that in the second example each subset is finite yet the union of all subsets contains *all* positive integers and hence is infinite, therefore cannot be a member of the family of sets.

Theorem 5.1. Let $\mathscr{F}$ be an inductive family of sets. Let f be a function on $\mathscr{F}$ such that if $A \in \mathscr{F}$, then $f(A) \in \mathscr{F}$ and $f(A) \subset A$. Then there is a set $A_0 \in \mathscr{F}$ such that $f(A_0) = A_0$.

As a preliminary example, consider the inductive family

$$\mathscr{F} = \{\,\{a,b\},\ \{a,b,c\}\,\}$$

and define
$$f(\{a,b\}) = \{a,b,c\}$$

$$f(\{a,b,c\}) = \{a,b,c\}$$

Therefore, the theorem is satisfied trivially. It can be observed that this will always be the case when $\mathscr{F}$ is a family consisting of a *finite* number of sets.

Proof: Let $K \in \mathscr{F}$. Then a family $\mathscr{G}$ is a K-support if:

(1) $K \in \mathscr{G}$

(2) $A \in \mathscr{G} \Rightarrow f(A) \in \mathscr{G}$

(3) $\mathscr{C} \subset \mathscr{G}$ and $\mathscr{C}$ is a chain $\Rightarrow \left(\bigcup_{A \in \mathscr{C}} A \right) \in \mathscr{G}$,

i.e. $\mathscr{G}$ is inductive.

Note that $\mathscr{F}$ is a K-support.

Now let $\Gamma = \{\mathscr{G} \mid \mathscr{G}$ is a K-support$\}$, and define $\mathscr{H} = \bigcap_{\mathscr{G} \in \Gamma} \mathscr{G}$ This means that $\mathscr{H}$ is a sub-family of every K support. An alternative way of defining $\mathscr{H}$ are:

$$\mathscr{H} = \{A \mid \mathscr{G} \text{ is a } K \text{ support} \Rightarrow A \in \mathscr{G}\}$$

or $\qquad \mathscr{H} = \{A \mid A \text{ belongs to every } K\text{-support}\}$

At this point, the completion of the proof rests upon two assertions about $\mathscr{H}$. These are:

(a) $\mathscr{H}$ is a K-support,

(b) $\mathscr{H}$ is a chain.

The procedure now will be to assume these two assertions and complete the proof on this basis. Then, a series of 6 lemmas will be proved, the 1st of which proves $\mathscr{H}$ is a K-support and the 6th that $\mathscr{H}$ is a chain.

Therefore, assuming $\mathscr{H}$ is a K-support and a chain, let $A_0 = \bigcup_{A \in \mathscr{H}} A$. Now $A_0 \in \mathscr{F}$ since $\mathscr{H}$ is a chain and $\mathscr{H} \subset \mathscr{F}$.

Therefore $f(A_0) \supset A_0$ by hypothesis. But since $\mathscr{H}$ is a chain and a K-support and $\mathscr{H} \subset \mathscr{H}$, then $\left(\bigcup_{A \in \mathscr{H}} A \right) \in \mathscr{H}$. In other words $A_0 \in \mathscr{H}$. Since $\mathscr{H}$ is a K-support, $f(A_0) \in \mathscr{H}$. $\therefore f(A_0) \subset \bigcup_{A \in \mathscr{H}} A$, i.e.

$$f(A_0) \subset A_0$$

But $f(A_0) \supset A_0$ and $f(A_0) \subset A_0$. Hence, $f(A_0) = A_0$, which is the conclusion of the theorem.

Now the remaining task is to prove the assertions (a), (b).

Lemma I. $\mathscr{H}$ is a K-support.

Proof: In order to show that $\mathscr{H}$ is a K-support, it must be shown that $\mathscr{H}$ has the three defining properties of a K-support.

(1) K belongs to every K-support, so

$$K \in \mathscr{H}$$

(2) If $A \in \mathscr{H}$, then A belongs to every K-support. Therefore $f(A)$ belongs to every K-support, by the definition of a K-support.

$$f(A) \in \mathscr{H}; \text{ i.e. } A \in \mathscr{H} \Rightarrow f(A) \in \mathscr{H}.$$

(3) If $\mathscr{C} \subset \mathscr{H}$ and $\mathscr{C}$ is a chain, then $\mathscr{C} \subset \mathscr{G}$ for every K-support $\mathscr{G}$.
Then $\left[\bigcup_{A \in \mathscr{C}} A\right] \in \mathscr{G}$ for every K-support $\mathscr{G}$.

$$\therefore \left[\bigcup_{A \in \mathscr{C}} A\right] \in \mathscr{H}; \text{ i.e. } \mathscr{H} \text{ is inductive.}$$

This completes the proof of Lemma 1 since $\mathscr{H}$ satisfies all of the conditions required for a family to be a K-support.

Lemma 2. (The principle of induction for $\mathscr{H}$.) Let $\mathscr{P} \subset \mathscr{H}$ such that

(1) $K \in \mathscr{P}$,

(2) $A \in \mathscr{P} \Rightarrow f(A) \in \mathscr{P}$,

(3) $\mathscr{C}$ is a chain and $\mathscr{C} \subset \mathscr{P} \Rightarrow \left[\bigcup_{A \in \mathscr{C}} A\right] \in \mathscr{P}$.

Then $\mathscr{P} = \mathscr{H}$.

Proof: By the definition of a K-support, $\mathscr{P}$ is a K-support. Then $\mathscr{H} \subset \mathscr{P}$, since by definition $\mathscr{H}$ is a sub-family of all K-supports. But $\mathscr{P} \subset \mathscr{H}$ by hypothesis. Therefore $\mathscr{P} = \mathscr{H}$.

Note. The usual principle of mathematical induction is quite analogous to this lemma. It can be stated as follows:

Let N be the set of all positive integers and let $S \subset N$ such that:

(1) $1 \in S$,

(2) $x \in S \Rightarrow x+1 \in S$.

Then, $S = N$.

Lemma 3. $A \in \mathscr{H} \Rightarrow A \supset K$

Proof: Let $\mathscr{M} = \{A \in \mathscr{H} \mid A \supset K\}$.

$$\therefore \mathscr{M} \supset \mathscr{H}$$

(1) $K \in \mathscr{H}$ and $K \supset K$.

$$\therefore K \in \mathscr{M}$$

(2) If $A \in \mathscr{M}$ then $A \supset K$. But $f(A) \supset A$ since $A \in \mathscr{H} \subset \mathscr{F}$. Therefore $f(A) \supset K$. Also $f(A) \in \mathscr{H}$ since $\mathscr{H}$ is a K-support. Therefore $f(A) \in \mathscr{M}$, i.e. $A \in \mathscr{M} \Rightarrow f(A) \in \mathscr{M}$.

(3) Let $\mathscr{C}$ be a chain where $\mathscr{C} \subset \mathscr{M}$, then $A \in \mathscr{C} \Rightarrow A \supset K$.

$$\therefore \left[\bigcup_{A \in \mathscr{C}} A \right] \supset K.$$

But $\left[\bigcup_{A \in \mathscr{C}} A \right] \in \mathscr{H}$ since $\mathscr{H}$ is a K-support.

$$\therefore \left[\bigcup_{A \in \mathscr{C}} A \right] \in \mathscr{M}$$

Then $\mathscr{M} = \mathscr{H}$ by Lemma 2, or $\mathscr{H} = \{ A \in \mathscr{H} \mid A \supset K \}$.

Therefore $A \in \mathscr{H} \Rightarrow A \supset K$ which is the conclusion of this lemma.

Definition: Let

$$\mathscr{L} = \{ A \in \mathscr{H} \mid [B \in \mathscr{H} \text{ and } B \neq A \text{ and } B \subset A] \Rightarrow f(B) \subset A \}$$

Thus, $\mathscr{L} \subset \mathscr{H}$.

Lemma 4. $A \in \mathscr{L}$ and $B \in \mathscr{H} \Rightarrow B \subset A$ or $B \supset f(A)$.

Proof. Let $A_0 \in \mathscr{L}$. Now define

$$\mathscr{N} = \{ B \in \mathscr{H} \mid B \subset A_0 \text{ or } B \supset f(A_0) \}$$

$$\therefore \mathscr{N} \subset \mathscr{H}$$

(1) $K \in \mathscr{H}$. Since $A_0 \in \mathscr{H}$, $K \subset A_0$ by Lemma 3.

$$\therefore K \in \mathscr{N}$$

(2) If $B \in \mathscr{N}$, does $f(B) \in \mathscr{N}$? First of all $f(B) \in \mathscr{H}$ since $\mathscr{H}$ is a K-support. There are now 3 cases to consider. $B \subset A_0$ gives

(a) $B = A_0$ or, (b) $B \subset A_0$ and $B \neq A_0$,

and the other case is   (c) $B \supset f(A_0)$

(a) If $B = A_0$	(b) If $B \subset A_0$	(c) $B \supset f(A_0)$
$f(B) = f(A_0)$	and $B \neq A_0$	$f(B) \supset B$
$\therefore f(B) \supset f(A_0)$	then $f(B) \subset A_0$	since $B \in \mathscr{F}$
$\therefore f(B) \in \mathscr{N}$	since $A_0 \in \mathscr{L}$	$\therefore f(B) \supset f(A_0)$
	$\therefore f(B) \in \mathscr{N}$	$\therefore f(B) \in \mathscr{N}$

Therefore $B \in \mathscr{N} \Rightarrow f(B) \in \mathscr{N}$.

(3) If $\mathscr{C}$ is a chain and $\mathscr{C} \supset \mathscr{N}$, does $\left[\bigcup_{B \in \mathscr{C}} B \right] \in \mathscr{N}$? Note that $\left[\bigcup_{B \in \mathscr{C}} B \right] \in \mathscr{H}$ since $\mathscr{H}$ is a K-support.

There are two cases to consider:

(a) For every $B \in \mathscr{C}$, it is the case that $B \subset A_0$. In this case

$$\left[\bigcup_{B \in \mathscr{C}} B \right] \subset A_0. \quad \therefore \quad \left[\bigcup_{B \in \mathscr{C}} B \right] \in \mathscr{N}$$

(b) There is some $B \in \mathscr{C}$ such that $B \subset A_0$ is false. But $B \in \mathscr{N}$. Therefore $B \supset f(A_0)$. However, $\left[\bigcup_{B \in \mathscr{C}} B \right] \supset B$.

$$\therefore \quad \left[\bigcup_{B \in \mathscr{C}} B \right] \supset f(A_0) \quad \text{or} \quad \left[\bigcup_{B \in \mathscr{C}} B \right] \in \mathscr{N}$$

Therefore, $\mathscr{C}$ is a chain and $\mathscr{C} \subset \mathscr{N} \Rightarrow \left[\bigcup_{B \in \mathscr{C}} B \right] \in \mathscr{N}$. From Lemma 2, the conclusion is that $\mathscr{N} = \mathscr{H}$. In other words,

$$\mathscr{H} = \{ B \in \mathscr{H} \mid B \subset A_0 \quad \text{or} \quad B \supset f(A_p) \}$$

when $A_0 \in \mathscr{L}$. Therefore $A \in \mathscr{L}$ and $B \in \mathscr{H} \Rightarrow B \subset A$ or $B \supset f(A_0)$ which is the conclusion of this lemma.

Lemma 5. $\mathscr{L} = \mathscr{H}$.

Proof: (1) $K \in \mathscr{L}$ is true vacuously because by Lemma 3, if $B \in \mathscr{H}$ then $B \supset K$. Hence the hypothesis is never satisfied.

(2) Does $A \in \mathscr{L} \Rightarrow f(A) \in \mathscr{L}$? That is, does $B \in \mathscr{H}$ and $B \neq f(A)$ and $B \subset f(A) \Rightarrow f(B) \subset f(A)$? The situation now is $B \in \mathscr{H}$ and $A \in \mathscr{L}$. Hence, from Lemma 4, $B \subset A$ or $B \supset f(A)$. Note that $f(A) \in \mathscr{H}$ since $\mathscr{H}$ is a K-support. Again we have the 3 cases to consider:

(a) $B = A$
  $f(B) = f(A)$
  $\therefore f(B) \subset f(A)$
  $\therefore f(A) \in \mathscr{L}$

(b) $B \subset A$ and $B \neq A$, then $f(B) \subset A$ since $A \in \mathscr{L}$ and B satisfies the conditions in the definition of $\mathscr{L}$. Also, $A \subset f(A)$ since $A \in \mathscr{F}$.
  $\therefore f(B) \subset f(A)$ or $f(A) \in \mathscr{L}$.

(c) $B \supset f(A)$.

But $B \subset f(A)$ by hypothesis.

Therefore $B = f(A)$ but this is impossible since $B \neq f(A)$ by hypothesis. Therefore, this case does not satisfy the hypothesis and the assertion is true vacuously.

Therefore $f(A) \in \mathscr{L}$. Then in all cases $A \in \mathscr{L} \Rightarrow f(A) \in \mathscr{L}$.

(3) If $\mathscr{C}$ is a chain and $\mathscr{C} \subset \mathscr{L}$, does $T = \left[\bigcup_{A \in \mathscr{C}} A \right] \in \mathscr{L}$? Note that $T \in \mathscr{H}$ since $\mathscr{H}$ is a K-support. Suppose there exist $B \in \mathscr{H}$ such that $B \neq T$ and $B \subset T$. Since $A \in \mathscr{C} \subset \mathscr{L}$, $B \subset A$ or $B \supset f(A) \supset A$, since $A \in \mathscr{F}$.

(a) Can it be that for all $A \in \mathscr{C}$, $B \supset A$? If so, $B \supset \bigcup_{A \in \mathscr{C}} A$, i.e. $B \supset T$. But $B \neq T$ and $B \subset T$ by hypothesis. Therefore this situation is impossible and $f(B) \subset T$ is true vacuously. Therefore $T \in \mathscr{L}$.

(b) There is some $A_0 \in \mathscr{C}$ such that $B \subset A_0$ and $B \neq A_0$. Now since $A_0 \in \mathscr{L}$ and $B \in \mathscr{H}$, then $f(B) \subset A_0$ since B satisfies the conditions in the definition of $\mathscr{L}$. But $A_0 \subset \bigcup_{A \in \mathscr{C}} A$, i.e. $A_0 \subset T$. Therefore $f(B) \subset T$ and $T \in \mathscr{L}$.

Then in all cases, $\mathscr{C}$ is a chain and $\mathscr{C} \subset \mathscr{L} \Rightarrow \left[\bigcup_{A \in \mathscr{C}} A \right] \in \mathscr{L}$. The conclusion now follows at once from Lemma 2.

Lemma 6. $\mathscr{H}$ is a chain.

Proof: Let A, $B \in \mathscr{H}$. Then $A \in \mathscr{L}$ since $\mathscr{L} = \mathscr{H}$ by Lemma 5. Therefore by Lemma 4, either $B \subset A$ or $B \supset f(A) \supset A$. Hence, $\mathscr{H}$ is a chain.

Definition 5.3. M is called a *maximal element* of a family $\mathscr{F}$ if:

$$A \in \mathscr{F} \quad \text{and} \quad A \supset M \Rightarrow A = M$$

Example:

Let $\mathscr{F} = \{\ \{1\},\ \{1, 2\},\ \{4\},\ \{4, 5\}\ \}$. $\mathscr{F}$ has the maximal elements $\{1, 2\}$ and $\{4, 5\}$.

Theorem 5.2. (Zorn's Lemma.) Let $\mathscr{F}$ be an inductive family. Then, $\mathscr{F}$ has a maximal element.

Proof (by contradiction): Suppose $\mathscr{F}$ has no maximal element. Then for each $A \in \mathscr{F}$ there is a $B \in \mathscr{F}$ such that: $A \subset B$ and $A \neq B$. For each $A \in \mathscr{F}$, let $T_A = \{B \in \mathscr{F} \mid B \supset A \text{ and } B \neq A\}$. Then for each $A \in \mathscr{F}$, $T_A \neq \emptyset$. Let $\Gamma = \{T_A \mid A \in \mathscr{F}\}$. By Theorem 4.1 (Principle of choice), there is a function g defined on Γ such that for each T_A, we have $g(T_A) \in T_A$. Define f so that $f(A) = g(T_A)$; so $f(A) \in T_A$. I.e., $f(A) \supset A$, $f(A) \neq A$. But, this is a contradiction, by Theorem 5.1.

6. The Functional Equation

$$f(x+y) = f(x) + f(y)$$

In this section all functions are real-valued functions of a real variable.

Example: Let $f(x) = kx$.

$$f(x+y) = k(x+y) = kx + ky = f(x) + f(y)$$

We will find all functions f that satisfy this equation.

Theorem 6.1. If f is defined for all real x, and if

$$f(x+y) = f(x) + f(y)$$

then $f(x) = f(1) \cdot x$ for all rational numbers x.

Proof:

$$f(x_1 + x_2 + \ldots + x_n) = f(x_1) + f(x_2) + \ldots + f(x_n)$$

Thus if n is a positive integer

$$\begin{aligned}
f(n) &= f(1 + 1 + 1 + \ldots + 1) \quad (n \text{ 1's}) \\
&= f(1) + f(1) + \ldots + f(1) \quad (n \, f(1)\text{'s}) \\
&= f(1) \cdot n
\end{aligned}$$

Finally, for $-n$ where n is a positive integer

$$0 = f(0) = f(n + (-n)) = f(n) + f(-n)$$

$$f(-n) = -f(n) = -f(1) \cdot n = f(1) \cdot (-n)$$

Next,

$$\begin{aligned}
f(0) &= f(1 + (-1)) \\
&= f(1) + f(-1) \\
&= f(1) - f(1) \\
&= 0 \\
&= 0 \cdot f(1)
\end{aligned}$$

Thus the result has been demonstrated for all integers x. For x rational, we have $x = m/n$ where n is positive. Hence,

$$nf\left(\frac{m}{n}\right) = f\left(\frac{m}{n}\right) + f\left(\frac{m}{n}\right) + \ldots + f\left(\frac{m}{n}\right) \qquad \left(n \qquad f\left(\frac{m}{n}\right)\text{'s}\right)$$

$$= f\left(\frac{m}{n} + \frac{m}{n} + \ldots + \frac{m}{n}\right) \qquad \left(n \qquad \frac{m}{n}\text{'s}\right)$$

$$= f(m) = f(1) \cdot m;$$

so $$f\left(\frac{m}{n}\right) = f(1) \cdot \frac{m}{n}$$

Theorem 6.2. If $f(x+y) = f(x) + f(y)$ and f is continuous, then $f(x) = f(1) \cdot x$ for all x.

Proof: Let x_0 be any real number. Let $x_0 = \lim_{n \to \infty} r_n$ where each r_n is rational then,

$$f(x_0) = \lim_{x \to x_0} f(x) = \lim_{n \to \infty} f(r_n)$$

$$= \lim_{n \to \infty} (f(1) \cdot r_n) = f(1) \lim_{n \to \infty} r_n$$

$$= f(1) \cdot x_0$$

In order to prove the existence of discontinuous solutions of the functional equation $f(x+y) = f(x) + f(y)$, we require the notion of Hamel basis.

Definition 6.1. A set Γ of real numbers is called a *Hamel basis* if $1 \in \Gamma$ and for each real number x there are *uniquely* determined numbers $x_1, x_2, \ldots, x_n \in \Gamma$ and non-zero rational numbers $r_1, r_2, \ldots, r_n$ such that:

$$x = r_1 x_1 + r_2 x_2 + \ldots + r_n x_n$$

Theorem 6.3. There is a Hamel basis.

Proof: A set of real numbers Γ will be called *nice* if:

(1) $1 \in \Gamma$,

(2) $r_1, r_2, \ldots, r_n$ rational and $x_1, x_2, \ldots, x_n \in \Gamma$ and x_i's all different and

$$\sum_{i=1}^{n} r_i x_i = 0 \Rightarrow r_1, r_2, \ldots, r_n = 0$$

Let $\mathscr{F}$ be the family:

$$\mathscr{F} = \{\Gamma \,|\, \Gamma \text{ is nice}\}$$

Lemma 1. Let $\mathscr{C}$ be a chain. Let $A_1, A_2, \ldots, A_n \in \mathscr{C}$. Then,

$$A_1 \cup A_2 \cup \ldots \cup A_n \in \mathscr{C}$$

Or, the union of a finite number of elements of a chain belongs to the chain.

Proof: The proof will be by mathematical induction on n. For $n = 1$ it is obvious—the union is just the one set. Suppose the result is true for $n = k$. Then for $n = k+1$,

let $\quad A_1, A_2, \ldots A_k, A_{k+1} \in \mathscr{C}$,

let $\quad R = A_1 \cup A_2 \cup \ldots \cup A_k$,

and let $S = R \cup A_{k+1}$

To prove: $S \in \mathscr{C}$.

$R \in \mathscr{C}$ by induction hypothesis.

$A_{k+1} \in \mathscr{C}$.

$R \subset A_{k+1}$ or $A_{k+1} \subset R$ since $\mathscr{C}$ is a chain.

If $R \subset A_{k+1}$ then $S = A_{k+1}$ and $S \in \mathscr{C}$.

If $A_{k+1} \subset R$ then $S = R$ and $S \in \mathscr{C}$.

Lemma 2. $\mathscr{F}$ is inductive.

Proof: Let $\mathscr{C} \subset \mathscr{F}$, where $\mathscr{C}$ is a chain. Let $T = \bigcup_{A \in \mathscr{C}} A$. To prove that $T \in \mathscr{F}$ or that T is a nice set:

(1) $1 \in T$ is obvious because: $1 \in A$ for each $A \in \mathscr{C}$.

(2) Let $r_1, r_2, \ldots, r_n$ be rational. Let $x_1, x_2, \ldots, x_n \in T$, x_i's all different. Let $\sum_{i=1}^{n} r_i x_i = 0$. To prove that the r_i's $= 0$:

Each $x_i \in T$.

Therefore $x_i \in A_i \in \mathscr{C}$, $i = 1, 2, \ldots, n$

Let $M = A_1 \cup A_2 \cup \ldots \cup A_n$.

Then $x_i \in M$, $i = 1, 2, \ldots, n$.

$M \in \mathscr{C}$ (By Lemma 1).

$\mathscr{C} \subset \mathscr{F}$.

$M \in \mathscr{F}$.

Thus, M is nice.

Therefore $r_1, r_2, \ldots, r_n = 0$, so that T is nice. This completes the proof of the lemma.

Since $\mathscr{F}$ is inductive, Zorn's lemma now tells us that $\mathscr{F}$ has a maximal element. Call this H. (We want to prove H is a Hamel basis.)

Suppose there is some ξ such that:

$$\xi \neq \sum_{i=1}^{n} r_i x_i, \quad x_i \in H, \quad \xi \notin H$$

Let $H' = H \cup \{\xi\}$. Then $H \subset H'$.

Claim: H' is nice.

(1) $1 \in H$. Therefore $1 \in H'$.

(2) Let $\sum_{i=1}^{n} r_i x_i = 0$, x_i's different, $x_i \in H'$. Suppose ξ is any one of the x's say x_1. Then

$$\xi = -\frac{r_2}{r_1} x_2 - \frac{r_3}{r_1} x_3 - \ldots - \frac{r_n}{r_1} x_n \quad \text{or} \quad \xi = \sum_{i=1}^{n} \left[-\frac{r_i}{r_1} x_i \right]$$

and this is impossible. So ξ is not any one of the x's.

Therefore $x_1, x_2, \ldots, x_n \in H$.

But, H is nice.

Therefore $r_1, r_2, \ldots, r_n = 0$.

Therefore H' is nice.

Therefore $H' \in \mathscr{F}$.

But H is a maximal element, and

$H \subset H'$.

$H' \in \mathscr{F}$.

Then $H = H'$.

But $\xi \notin H$ and $\xi \in H'$. Therefore for each ξ, $\xi = \sum_{i=1}^{n} r_i x_i$, x_i's all different and $x_i \in H$.

It remains only to prove that this representation is unique. Suppose. for $x_1, \ldots, x_n \in H$,

$$x = r_1 x_1 = r_2 x_2 + \ldots + r_n x_n$$
$$x = s_1 x_1 + s_2 x_2 + \ldots + s_n x_n$$

Here some of the r_i's and s_i's may be 0. Then, subtracting:

$$0 = (r_1 - s_1)x_1 + (r_2 - s_2)x_2 + \ldots + (r_n - s_n)x_n$$

Since H is nice, these coefficients are all $= 0$. I.e., $r_1 = s_1, r_2 = s_2, \ldots, r_n = s_n$.

Theorem 6.4. Let H be any Hamel basis. Let $f(x)$ be defined arbitrarily for $x \in H$ and let

$$f(x) = \sum_{i=1}^{n} r_i f(x_i) \quad \text{for} \quad x = \sum_{i=1}^{n} r_i x_i, \quad x_i \in H$$

Then, $\qquad\qquad\qquad f(x+y) = f(x) + f(y)$

Proof: Let

$$x = r_1 x_1 + r_2 x_2 + \ldots + r_n x_n$$
$$y = s_1 x_1 + s_2 x_2 + \ldots + s_n x_n$$

Then, $\qquad f(x) = \sum r_i f(x_i)$

and $\qquad f(y) = \sum s_i f(x_i)$

Now, $\qquad x + y = (r_1 + s_1)x_1 + (r_2 + s_2)x_2 + \ldots + (r_n + s_n)x_n$

and therefore

$$f(x+y) = \sum (r_i + s_i)f(x_i)$$

Thus, $\qquad f(x+y) = f(x) + f(y)$

Problems

1. Which of the following statements are true and which are false?

   (a) $\{1, 2, 3\} \subset \{1, 2, 3, 4\}$.

   (b) $\{1\} \in (1, 2)$.

   (c) $\{1, 2, 3\} \cap \{3, 4, 5\} = 3$.

   (d) $\{1, 2\} \cup \{2, 3\} = \{1, 2, 3\}$.

   (e) $\{1\} \in 2\{1, 2, 3\}$.

   (f) $(1, 2) \in \{1, 2, 3\} \times \{1, 2, 3\}$.

2. Let:

> j be a certain definite dog named Jimmy,
>
> D be the class of all dogs,
>
> S be the set of all species of animals, and
>
> A be the set of all animals.

(a) Write all true statements of the forms $x \in y$ and $x \subset y$ where x and y can be j, D, S, or A.

(b) Find a solution in the set $\{j, D, S, A\}$ for the equation: $2^x = y$.

3. Prove using the definition of ordered pair that $(x, y) = (x', y')$ implies $x = x'$ and $y = y'$.

4. (a) For each of the following relations tell whether it is (i) reflexive, (ii) symmetric, and (iii) transitive:

   (1) $|x| = |y|$, x, y complex numbers.

   (2) $x < y$, x, y real numbers.

   (3) $x^y = y^x$, $x > 0$, $y > 0$.

   (4) $x \subset y$, x, y *sets of* real numbers.

(b) Which are equivalence relations?

(c) For those which are, describe the equivalence classes.

5. (a) Which of the following families are inductive?

   (1) The family of all *finite* sets of positive integers.

   (2) The family of all sets of positive integers.

   (3) The family of all sets which are residue classes of positive integers modulo some integer. (E.g. the set of even positive integers, and the set of multiples of 3, but *not* the set of perfect squares, belongs to this family.)

(b) Choosing one example in (a) which is inductive, show how the proof of Theorem 5.1 would work out. (You may ignore the Lemmas.)

6. Let Q be some definite set of real-valued functions defined for $-1 \leq x \leq 1$. Let,

$$\mathscr{F}_Q = \{\{f, -f\} \mid f \in Q\}$$

(E.g. if $f(x) = x^2$ is in Q, then the set $\{f, g\} \in \mathscr{F}_Q$ where

$$g(x) = -x^2.)$$

By the multiplicative principle, there is a set R such that for each $f \in Q$, either $f \in R$ or $-f \in R$ but not both. Can you define a set R with this property, *but not using* the multiplicative principle, if:

(a) Q is the set of linear functions?

(b) Q is the set of polynomials?

(c) Q is the set of functions expressible as convergent power series

$$\sum_{n=0}^{\infty} a_n x^n \quad \text{for} \quad |x| < 1$$

(d) Q is the set of functions continuous for $-1 \leq x \leq 1$.

(e) Q is the set of *all* functions defined for $-1 \leq x \leq 1$.

(f) Q is the set of solutions of $f(x+y) = f(x) + f(y)$.

(This problem was communicated by Dr. Marvin Minsky, who attributed it to J. von Neumann.)

CHAPTER 2

Normed Linear Spaces and Algebras

I. Definitions

We shall let R denote the real number field, C denote the complex number field, and F denote either R or C.

Definition 1.1. A *linear space* over F is a set, L, taken together with a function, $+$, from $L \times L$ into L, and a function $\cdot$ from $F \times L$ into L such that

(1) $X, Y \in L \Rightarrow X + Y = Y + X$,

(2) $X, Y, Z \in L \Rightarrow X + (Y + Z) = (X + Y) + Z$,

(3) there is an element 0 in L such that $X + 0 = X$,

(4) $X \in L \Rightarrow \exists (-X) \in L$ such that $X + (-X) = 0$.

((1) − (4) means that L, with the operation $+$, forms an Abelian Group.)

(5) $a, b \in F$ and $X \in L \Rightarrow a(bX) = (ab)X$,

(6) $a \in F$ and $X, Y \in L \Rightarrow a(X + Y) = aX + aY$,

(7) $a, b \in F$ and $X \in L \Rightarrow (a + b)X = aX + bX$,

(8) $X \in L \Rightarrow 1 \cdot X = X$,

(9) $X \in L \Rightarrow 0 \cdot X = 0$.

The dot $\cdot$ will usually be omitted.

Definition 1.2. L is a *normed linear space* over F if L is a linear space over F and for each $X \in L$ there is a number $\| X \|$ such that:

(1) $\| X \| \geq 0$,

(2) $\| X + Y \| \leq \| X \| + \| Y \|$,

(3) $\| aX \| = |a| \| X \|$,

(4) $\| X \| = 0 \Leftrightarrow X = 0$.

23

Definition 1.3. A *sequence* of elements of a normed linear space is a function from the positive integers into L.

If X is a sequence we write $X_n = X(n)$ and $X = \{X_n\}$

Definition 1.4. $X_n \to X$, or $\lim_{n \to \infty} X_n = X$ means $\lim_{n \to \infty} \| X_n - X \| = 0$.

Corollary 1.1. If $X_n \to X$ and $X_n \to X'$, then $X = X'$.

Proof:
$$\| X - X' \| = \| X_n - X' - X_n + X \| = \| X_n - X' - 1 \cdot X_n + 1 \cdot X \|$$
$$= \| X_n - X' + (-1)X_n - (-1)X \|$$
$$= \| (X_n - X') + (-1)(X_n - X) \|$$
$$\leq \| X_n - X' \| + \| (-1)(X_n - X) \|$$
$$\| X - X' \| \leq \| X_n - X' \| + \| X_n - X \|$$

But, $\| X_n - X' \| \to 0; \quad \| X_n - X \| \to 0$

Thus, $\| X - X' \| = 0, \quad$ i.e. $X = X'$

Corollary 1.2. $X_n \to X$ and $Y_n \to Y \Rightarrow aX_n + bY_n \to aX + bY$.

Proof:
$$\| (aX_n + bY_n) - (aX + bY) \| = \| a(X_n - X) + b(Y_n - Y) \|$$
$$\leq \| a(X_n - X) \| + \| b(Y_n - Y) \|$$
$$= |a| \cdot \| X_n - X \| + |b| \cdot \| Y_n - Y \| \to 0$$

Thus, $\| (aX_n + bY_n) - (aX + bY) \| \to 0$

Therefore, $aX_n + bY_n \to aX + bY$

Definition 1.5. $\{X_n\}$ is a *Cauchy sequence* if $\lim_{\substack{m \to \infty \\ n \to \infty}} \| X_m - X_n \| = 0$;

that is, if for every $\epsilon > 0$, there is an N such that:
$$m, n > N \Rightarrow \| X_m - X_n \| < \epsilon$$

Definition 1.6. $\{X_n\}$ *converges* if there exists an X such that $X_n \to X$.

Corollary 1.3. If $\{X_n\}$ converges, then $\{X_n\}$ is a Cauchy sequence.

Proof: Let $X_n \to X$.

Consider

$$\| X_n - X_m \| = \| X_n - X - X_m + X \| = \| (X_n - X) - (X_m - X) \|$$
$$\leq \| X_n - X \| + \| X_m - X \| \to 0$$

as $m, n \to \infty$.

Therefore, $\lim_{\substack{m \to \infty \\ n \to \infty}} \| X_n - X_m \| = 0$, i.e. $\{X_n\}$ is a Cauchy sequence.

Definition I.7. A normed linear space L is called *complete* if every Cauchy sequence in it converges.

A complete normed linear space is also called a *Banach Space*.

Several examples of Banach spaces will be studied later. In fact, functional analysis, is largely concerned with the theory of Banach spaces.

Definition I.8. $\sum_{n=1}^{\infty} X_n = X$ means $Y_n \to X$, where

$$Y_n = \sum_{k=1}^{n} X_k = X_1 + X_2 + \ldots + X_n$$

$\sum_{n=1}^{\infty} X_n$ *converges* means that $\sum_{n=1}^{\infty} X_n = X$ for some X.

Corollary I.4. In a Banach space, if $\sum_{n=1}^{\infty} \| X_n \|$ converges, then $\sum_{n=1}^{\infty} X_n$ converges.

Proof: Let $Y_n = \sum_{k=1}^{n} X_k$ and let $t_n = \sum_{k=1}^{n} \| X_k \|$. Now we show that the terms of the sequence $\| Y_m - Y_n \|$ can be made small. Without loss of generality, let $m > n$. Then,

$$\| Y_m - Y_n \| = \| \sum_{k=1}^{m} X_k - \sum_{k=1}^{n} X_k \|$$
$$= \| X_{n+1} + X_{n+2} + X_{n+3} + \ldots + X_m \|$$
$$\leq \| X_{n+1} \| + \| X_{n+2} \| + \ldots + \| X_m \|$$
$$= | t_m - t_n | \to 0 \text{ as } m, n \to \infty$$

$\therefore \left\{ \sum_{k=1}^{n} X_k \right\}$ is a Cauchy sequence, and since this is a Banach space, $\sum_{n=1}^{\infty} X_n$ converges.

Definition 1.9. A linear space L over F is called a *linear algebra over F* if there is defined a function $\circ$ from $L \times L$ into L such that:

(1) $X, Y, Z \in L \Rightarrow X \circ (Y \circ Z) = X \circ Y) \circ Z$

(2) There is an element $e \in L$ such that
$$e \circ X = X \circ e = X$$

(3) $X \circ (Y+Z) = X \circ Y + X \circ Z$ and $(X+Y) \circ Z = X \circ Z + Y \circ Z$ for $X, Y, Z \in L$.

(4) $a(X \circ Y) = (aX) \circ Y$.

Definition 1.10. A linear algebra is called *Abelian* if $X \circ Y = Y \circ X$.

Definition 1.11. A *normed linear algebra* is a normed linear space which is also a linear algebra and such that:
$$\| e \| = 1,$$
and
$$\| X \circ Y \| \leq \| X \| \cdot \| Y \|$$

Note. The symbol $\circ$ will usually be written $\cdot$ and will often be omitted altogether.

Corollary 1.5. $X_n \to X \Rightarrow \| X_n \| \to \| X \|$

Proof:
$$\| X_n \| = \| X_n - X + X \|$$
$$\leq \| X_n - X \| + \| X \|$$

Thus
$$\| X_n \| - \| X \| \leq \| X_n - X \|$$

Applying the same process to $\| X \|$
$$\| X \| = \| X - X_n + X_n \|$$
$$\leq \| X - X_n \| + \| X_n \|$$
$$= \| X_n - X \| + \| X_n \|$$

Thus
$$\| X \| - \| X_n \| \leq \| X_n - X \|$$

That is to say
$$0 \leq | \| X_n \| - \| X \| | \leq \| X_n - X \| \to 0$$

Therefore
$$| \| X_n \| - \| X \| | \to 0, \text{ or } \| X_n \| \to \| X \|$$

Corollary 1.6. In a normed linear space:

$$X_n \to X \text{ and } Y_n \to Y \Rightarrow X_n Y_n \to XY$$

Proof:
$$
\begin{aligned}
\| X_n Y_n - XY \| &= \| X_n Y_n - X_n Y + X_n Y - XY \| \\
&= \| X_n (Y_n - Y) + (X_n - X) Y \| \\
&\leq \| X_n \| \cdot \| Y_n - Y \| + \| X_n - X \| \cdot \| Y \| \\
&\to 0
\end{aligned}
$$

Definition 1.12. A normed linear space which is complete is called a *Banach algebra*.

Definition 1.13. In a linear algebra, Y is:

(1) an *inverse* of X if $XY = YX = e$.

(2) a *right-inverse* of X if $XY = e$.

(3) a *left-inverse* of X if $XY = e$.

If Y and Z are both inverses of X, we have $Y = YXZ = eZ = Z$.

Convention. We write $Y = X^{-1}$ to mean that X has an inverse, and that its value is Y.

Theorem 1.7. In a Banach algebra:

$$\| e - X \| < 1 \Rightarrow X \text{ has an inverse}$$

The proof of this fact may be motivated by the following identity valid for real or complex numbers a for which $| 1 - a | < 1$:

$$a^{-1} = \frac{1}{a} = \frac{1}{1 - (1-a)} = 1 + (1-a) + (1-a)^2 + \ldots$$

Proof. Consider, then, the series $\sum_{n=0}^{\infty} (e-X)^n$, with the purpose first of showing that it converges. The series $\sum_{n=0}^{\infty} \| e - X \|^n$ converges, because it is simply a geometric series of real numbers and because $\| e - X \| < 1$ by hypothesis. Now

$$
\begin{aligned}
\| (e-X)^n \| &= \| (e-X) \cdot (e-X) \cdot \ldots \cdot (e-X) \| \\
&\leq \| e-X \| \cdot \| e-X \| \cdot \ldots \cdot \| e-X \| = \| e-X \|^n
\end{aligned}
$$

Thus the series $\sum\limits_{n=0}^{\infty} \| (e-X)^n \|$ converges, by the ordinary comparison

test. Therefore, by Corollary 1.4, $\sum\limits_{n=0}^{\infty} (e-X)^n$ converges. Let

$Y = \sum\limits_{n=0}^{\infty} (e-X)^n$, and let us show that it is an inverse of X.

$$YX = Y[e-(e-X)] = Y + Y[-(e-X)]$$
$$XY = [e-(e-X)]Y = eY + [-(e-X)]Y$$
$$= Y + [-(e-X)]Y$$

Remembering that $Y = \lim\limits_{n\to\infty} \sum\limits_{k=0}^{n} (e-X)^k$, let us compute

$$Y[-(e-X)] \quad \text{and} \quad [-(e-X)]Y$$

$$Y[-(e-X)] = \lim\limits_{n\to\infty} \sum\limits_{k=0}^{n} (e-X)^k \ (e-X)$$

$$= \lim\limits_{n\to\infty} \sum\limits_{k=0}^{n} -(e-X)^{k+1} = \lim\limits_{n\to\infty} \sum\limits_{k=1}^{n+1} -(e-X)^k$$

$$= - \sum\limits_{n=1}^{\infty} (e-X)^n = e-Y$$

$$[-(e-X)]Y = \lim\limits_{n\to\infty} \sum\limits_{k=0}^{n} -(e-X)^k$$

$$= \lim\limits_{n\to\infty} \sum\limits_{k=0}^{n} -(e-X)^{k+1} = \lim\limits_{n\to\infty} \sum\limits_{k=1}^{\infty} -(e-X)^k = e-Y$$

Therefore $YX = Y + (e-Y) = e$
$$XY = Y + (e-Y) = e$$

2. Topology in a Normed Linear Space

In this section, L is some fixed definite normed linear space.

Definition 2.1. $A \subset L$ is called *closed* if $X_n \in A$ for all n implies that and $X_n \to X \Rightarrow X \in A$

Definition 2.2. $A \subset L$ is called *open* if $X \in A \Rightarrow \exists \ \epsilon > 0$ such that $\{Y \in L \mid \| X-Y \| < \epsilon\} \subset A$.

Note. A set may be neither open nor closed.

Theorem 2.1. A is open $\Leftrightarrow L-A$ is closed.

Proof: $\Rightarrow$ Let $\{X_n\}$ be some sequence such that $X_n \in L-A$ where $X_n \to X$. To prove $X \in L-A$. Suppose that $X \in A$. Since A is open then $\exists\ \epsilon > 0$ such that $\{Y \in L \mid \parallel X-Y \parallel < \epsilon\} \subset A$. But,

$$\parallel X_n - X \parallel \to 0$$

Therefore, $\exists\ N$ such that $n > N \Rightarrow \parallel X_n - X \parallel < \epsilon \Rightarrow X_n \in A$. But for all n, $X_n \in L-A$ which is a contradiction.

Proof: $\Leftarrow$ Let $X \in A$. To prove $\exists\ \epsilon > 0$ such that

$$\{Y \in L \mid \parallel X-Y \parallel < \epsilon\} \subset A.$$

Suppose this is not true. Then, for every $\epsilon > 0$, $\exists\ Y \in L$ such that $\parallel X-Y \parallel < \epsilon$, but $Y \notin A$. In particular we may pick $\epsilon = 1/n$ Call the Y, Y_n. Therefore, $\parallel X-Y_n \parallel < 1/n$, but $Y_n \notin A$, i.e. $Y_n \in L-A$. Now, $Y_n \to X$. Hence, since $L-A$ is closed, $X \in L-A$, which is a contradiction.

Definition 2.3. Let $A \subset L$. Let $\mathscr{F} = \{B \subset L \mid B \supset A$ and B is closed$\}$. $\bar{A} = \bigcap_{B \in \mathscr{F}} B$. $\bar{A}$ is called the *closure* of A.

Corollary 2.2. $A \subset \bar{A}$.

Proof: $\bar{A} = \bigcap_{B \in \mathscr{F}} B$, but every $B \in \mathscr{F}$ includes A.

Corollary 2.3. $\bar{A}$ is closed.

Proof: Let $X_n \in \bar{A}$ and $X_n \to X$. To prove, $X \in \bar{A}$. Since $X_n \in \bar{A}$, for each closed $B \supset A$, $X_n \in B$, which implies that for each closed $B \supset A$, $X \in B$. Thus, by Definition 2.3 $X \in \bar{A}$.

Corollary 2.4. If $B \supset A$ and B is closed then $B \supset \bar{A}$.

Proof: Let $X \in \bar{A}$; then, by Definition 2.3, $\bar{A} = \bigcap_{B \in \mathscr{F}} B$. Hence, $X \in B$.

Definition 2.4. Interior $(A) = L - \overline{(L-A)}$.

Corollary 2.5.   If $X \in \bar{A}$, then there exists a sequence $X_n \in A$ such that $X_n \to X$.

Proof. Let B be the set defined by:

$B = \{X \in L \mid$ there exists a sequence $\{X_n\}$, such that $X_n \in A$, and $X_n \to X\}$.

First note that $A \subset B$. (For, if $X \in A$, then X is the limit of the sequence $X, X, X, \ldots$ so that $X \in B$.)

Next, we note that B is closed. For, let $Y_n \in B$, $Y_n \to Y$. Then we must show that $Y \in B$. Since $Y_n \in B$, there is a sequence, $X_m^{(n)} \in A$, such that $X_m^{(n)} \to Y_n$ as $m \to \infty$. Hence, for each n, there is an integer m_n such that

$$\| X_{m_n}^{(n)} - Y_n \| < \frac{1}{n}$$

Let $Z_n = X_{m_n}^{(n)}$. Then $Z_n \in A$, and

$$\| Z_n - Y \| \leq \| Z_n - Y_n \| + \| Y_n - Y_n \|$$

$$\leq \frac{1}{n} + \| Y_n - Y \| \to 0$$

Therefore $Y \in B$. Thus, B is closed. However by Corollary 2.4, $A \subset B$ and B closed implies $\bar{A} \subset B$. Hence $\bar{A} \subset B$.

Definition 2.4. $A \subset L$ is *bounded* means there exists an $M > 0$ such that

$$X \in A \Rightarrow \| X \| \leq M$$

This concept can be thought of in terms of a geometric picture. If the set can be enclosed in a sphere, it is bounded.

Definition 2.5. $A \subset L$ is called *compact* if whenever $X_n \in A$, there is a convergent subsequence of $\{X_n\}$ whose limit is in A.

Corollary 2.5.   A compact set is closed.

Proof: Let A be compact. Let $X_n \in A$, $X_n \to X$. Then it must be shown that $X \in A$.

This is obvious, since every subsequence of X_n approaches X. So $X \in A$.

Definition 2.6.   The sequence $\{X_n\}$ is *bounded* means that there exists an $M > 0$, such that $\| X_n \| \leq M$.

Corollary 2.7. A convergent sequence is bounded.

Proof: Let $X_n \to X$. Then there is an integer N such that
$$n > N \Rightarrow \| X_n - X \| < 1$$
Then
$$n > N \Rightarrow \| X_n \| = \| (X_n - X) + X \| \leq \| X_n - X \| + \| X \| < 1 + \| X \|$$
Now let $M = \max(\| X_1 \| , \| X_2 \| , \ldots , \| X_n \| , 1 + \| X \|)$.
Then, $\| X_n \| \leq M$.

Corollary 2.8. A compact set is bounded.

Proof: Let A be compact. Suppose A is unbounded. Then, for each positive integer n, there is an $X_n \in A$ such that
$$\| X_n \| > n \text{ (using principle of choice)}$$
Since A is compact, some subsequence
$$\{X_{n_k}\} \text{ converges}$$
Therefore $\| X_{n_k} \| \leq M$ (by Corollary 2.7).

However $\| X_{n_k} \| > n_k > M$ (for some value of k). And so, we arrive at a contradiction.

Definition 2.7. The space L is called *piecewise compact* if every closed, bounded set in L is compact.

Corollary 2.9. A Cauchy sequence is bounded.

Proof: Let $\{X_n\}$ be a Cauchy sequence. Then, there is an integer N such that:
$$m, n > N \Rightarrow \| X_n - X_m \| < 1$$
Hence, $\qquad n > N \Rightarrow \| X_n - X_{N+1} \| < 1$

The rest of the proof is identical to that of Corollary 2.7 with the role, played there by X, being played here by X_{N+1}.

Corollary 2.10. If A is bounded, then $\bar{A}$ is bounded.

Proof: Let $Y \in A \Rightarrow \| Y \| \leq M$. Let $X \in \bar{A}$. Then, by Corollary 2.5 there exists $X_n \in A$, $X_n \to X$. Now, $\| X_n \| \leq M$. Thus,
$$\lim_{n \to \infty} \| X_n \| \leq M, \text{ and so } \| X \| \leq M,$$
which completes the proof.

Theorem 2.11. If L is piecewise compact, then L is a Banach space.

Proof: Let $\{X_n\}$ be a Cauchy sequence. Then $\{X_n\}$ is bounded (Corollary 2.9). Let A be the set of all terms of $\{X_n\}$. Then, A is bounded. Hence $\bar{A}$ is bounded. (Corollary 2.10). Hence the closure of $\bar{A}$ is compact because $\bar{A}$ is closed (Corollary 2.3) and bounded. Therefore $\{X_n\}$ has a subsequence $\{X_{n_k}\}$ which converges to a point $X \in \bar{A}$ (Definition of compactness). Let $\epsilon > 0$ be given. Choose N such that $m, n > N \Rightarrow \| X_n - X_m \| < \epsilon$ (definition of Cauchy sequence). There is an integer k_0 such that $n_{k_0} > N$ and $\| X_{n_{k_0}} - X \| < \epsilon/2$ since $X_{n_k} \to X$. Thus, $n > N$ implies

$$\| X_n - X \| \leqq \| X_n - X_{n_{k_0}} \| + \| X_{n_{k_0}} - X \| < \tfrac{1}{2}\epsilon + \tfrac{1}{2}\epsilon = \epsilon.$$

Hence $X_n \to X$. This implies that L is complete, i.e. that it is a Banach space.

3. R as a Normed Linear Space

In this section, all elements are in R, $\| X \| = | X |$, so that R is a normed linear space over R.

Definition 3.1. M is called an *upper bound* of the set of real numbers A if $X \in A \Rightarrow X \leqq M$.

Theorem 3.1. If A has an upper bound, then there is exactly one number S such that:

(1) S is an upper bound of A.

(2) If $X < S$, then X is not an upper bound of A.

This theorem expresses the so-called "completeness" property of the real numbers, and we accept it here without proof.

Definition 3.1. We write $S = \sup A$ if S is the number of Theorem 3.1. S is called the supremum or least upper bound of A.

If A has no upper bound, we write $\sup A = \infty$.

Similarly we may define $S = \inf A$ where S is the greatest lower bound of A. Details are left to the student.

Definition 3.3. $\{X_n\}$ is *monotone increasing* if $X_n \leqq X_{n+1}$.

Theorem 3.2. If $\{X_n\}$ is monotone increasing and bounded, then it converges.

Proof: Let A be the set of all terms of $\{X_n\}$. Let $X = \sup A$. Then $X_n \leq X$. Let $\epsilon > 0$ be any number. Then $X - \epsilon$ is not an upper bound for A. Therefore, there exists an N such that $X_N > X - \epsilon$. Now, $n > N \Rightarrow X_n \geq X_N > X - \epsilon$. $n > N \Rightarrow X - \epsilon < X_n \leq X$.

$$n > N \Rightarrow -\epsilon < X_n - X \leq 0.$$

$n > N \Rightarrow |X_n - X| < \epsilon$. Thus, $X_n \to X$.

The definition of monotone decreasing sequence and the statement and proof of an analogue of Theorem 3.2 are left to the student.

Theorem 3.3. Every sequence has either a monotone increasing or a monotone decreasing subsequence.

For the purpose of proving this theorem we define: X_k is a *peak of the sequence* $\{X_n\}$ if $n > k \Rightarrow X_n \leq X_k$.

Lemma. If $\{X_n\}$ has no monotone increasing subsequences, then for each r_0, there is a $k > r_0$ such that X_k is a peak.

Proof: Suppose otherwise. Then there is some r_0 such that for no $k > r_0$ is X_k a peak. $X_{N_1} = X_{r_0+1}$ is not a peak. There is $N_2 > N_1$, $X_{N_2} > X_{N_1}$. Continuing this process we see that $\{X_n\}$ has the monotone increasing subsequence $X_{N_1}, X_{N_2}, \ldots$. This is a contradiction.

Proof of Theorem 3.3: Let $\{X_n\}$ be a sequence. Suppose $\{X_n\}$ doesn't have a monotone increasing subsequence. Then, (take $r = 1$ in lemma) $\exists\, k_1$ such that X_{k_1} is a peak. Next take $r = k_1$, $\exists\, k_2 > k_1$ such that X_{k_2} is a peak. There exist $k_1 < k_2 < k_3 < k_4 \ldots$ such that each X_{k_t} is a peak. Hence $X_{k_1} \geq X_{k_2} \geq X_{k_3} \geq \ldots$ is a monotone decreasing subsequence.

Theorem 3.4. (Bolzano-Weierstrass theorem) R is piecewise compact.

Proof: Let $A \subset R$ be bounded and closed. Let $X_n \in A$; $\{X_n\}$ has a monotone subsequence $\{X_{n_k}\}$ (Theorem 3.3). $\{X_{n_k}\}$ is bounded because A is bounded. Hence $\{X_{n_k}\}$ converges. (Bounded monotone sequences converge.) Say $X_{n_k} \to X$. But $X \in A$ since A is a closed set. Hence A is compact.

Corollary 3.5. (Cauchy convergence criterion.) R is a Banach space.

Proof: By Theorem 3.4, R is piecewise compact, and hence, by Theorem 2.11 it is a Banach space.

4. The Cartesian Product of Normed Linear Spaces

In this section L and Q are two normed linear spaces over the same field F.

The Cartesian product $L \times Q$ can be made into a normed linear space by defining for $X, X' \in L$, $Y, Y' \in Q$, and $a \in F$:

$$(X, Y)+(X', Y') = (X+X', y = Y')$$
$$a \cdot (X, Y) = (aX, aY)$$
$$\| (X, Y) \| = \sqrt{\| X \|^2 + \| Y \|^2}$$

It is left to the reader to show that $L \times Q$ is a normed linear space and that $\| X \| \leq \| (X, Y) \|$, $\| Y \| \leq \| (X, Y) \|$.

Theorem 4.1. If L and Q are Banach spaces, then so is $L \times Q$.

Proof: Let $\{(X_n, Y_n)\}$ be a Cauchy sequence in $L \times Q$.

$$\| X_n - X_m \| \leq \| (X_n, Y_n)-(X_m, Y_m) \|$$

Thus, $\{X_n\}$ is a Cauchy sequence. Similarly, $\{Y_n\}$ is a Cauchy sequence. Let $X_n \to X$, $Y_n \to Y$. Then,

$$\| (X_n, Y_n)-(X, Y) \| = \| (X_n-X, Y_n-Y) \|$$
$$= \sqrt{\| X_n-X \|^2 + \| Y_n-Y \|^2} \to 0$$

I.e., $(X_n, Y_n) \to (X, Y)$.

Corollary 4.2. C is a Banach space.

Proof: $C = R \times R$.

Theorem 4.3. If L and Q are piecewise compact, then so is $L \times Q$.

Proof: Let $A \subset L \times Q$ where A is bounded and closed. Let M be such that $(X, Y) \in A \Rightarrow \| (X, Y) \| \leq M$. Let $\{(X_n, Y_n)\}$ be a sequence such that $(X_n, Y_n) \in A$.

Consider $\{X_n\}$. Then $\| X_n \| \leq \| (X_n, Y_n) \| \leq M$. Let B be the set of terms of $\{X_n\}$. Then $\bar{B}$ is closed and bounded. Therefore $\bar{B}$ is compact. Therefore there exists a subsequence.

$$X_{n_k} \to X \in \bar{B}$$

Now let E be the set of terms of $\{Y_{n_k}\}$, the corresponding subsequence of $\{Y_n\}$. $\bar{E}$ is closed and bounded. Therefore $\bar{E}$ is compact. Therefore there exists a subsequence

$$Y_{n_{k_l}} \to \in \bar{E}$$

Therefore $(X_{n_{k_l}}, Y_{n_{k_l}}) \to (X, Y) \in A$ since A is closed.

Corollary 4.4. C is piecewise compact.

Proof: $C = R \times R$ and R is piecewise compact.

Notation

(1) The n dimensional Cartesian product

$$L_1 \times L_2 \times L_3 \times \ldots \times L_n = L_1 \times (L_2 \times (L_3 \times \ldots \times (L_{n-1} \times L_n)$$
$$\ldots)$$

(2) If each of the L_i above are the same, then we write

$$L^n = L \times L \times \ldots \times L,$$

to n factors.

Corollary 4.5. If L is piecewise compact then L^n is piecewise compact. If L is a Banach Space, then L^n is a Banach Space.

Proof: By induction:

(1) L^1 is piecewise compact by hypothesis.

(2) If L^n is piecewise compact, then $L^{n+1} = L \times L^n$ is piecewise compact by Theorem 4.3.

The same argument works for being a Banach space.

The space R^n is called n-dimensional Euclidean Space.

The space C^n is called n-dimensional Complex Euclidean Space.

Problems

1. Prove that in an Abelian Banach algebra:

$$(X+Y)^n = \sum_{k=0}^{n} {_nC_k} X^k Y^{n-k}$$

where n is a positive integer and

$$_nC_k = \frac{n!}{k!(n-k)!}$$

2 (a) Prove that in any Banach algebra the following series converge for all X:

$$e + X + \frac{1}{2!}X^2 + \frac{1}{3!}X^3 + \dots$$

(b) Calling the sum in (a), exp (X), prove that exp $(X+Y)$ = exp (X) exp (Y) in any Abelian Banach algebra. (You may assume without proof that multiplication of absolutely convergent series is legitimate in a Banach algebra.)

Definition for Problem 3

An *inner product space* I over R is defined as follows:

(1) I is a linear space over R.

(2) There is an operation (called "inner product") from $I \times I$ into R whose value for given $(X, Y) \in I \times I$ will be written $[X, Y]$, with the following properties:

(a) $[X_1 + X_2, Y] = [X_1, Y] + [X_2, Y]$

(b) $[X, Y] = [Y, X]$

(c) $[aX, Y] = a[X, Y]$

(d) $[X, X] \geq 0$

(e) $[X, X] = 0 \Rightarrow X = 0$.

3. (a) In an inner product space prove Schwarz's inequality:

$$| [X, Y] | \leq \sqrt{(X, X)} \sqrt{(Y, Y)}$$

(b) Show that with the definition $\| X \| = \sqrt{(X, X)}$ an inner product space becomes a normed linear space.

(c) Show that using the "dot product" of elementary vector analysis, R^3 may be regarded as an inner product space and that the definition in (b) gives the usual norm.

(d) What about R^n?

4. Prove that:

(a) $\bar{\bar{A}} = \bar{A}$

(b) A is closed $\Leftrightarrow A = \bar{A}$

(c) $\overline{A \cup B} = \bar{A} \cup \bar{B}$

(d) Interior (A) is open

(e) Prove that if A and B are closed, then so are $A \cup B$ and $A \cap B$.

5. (a) Define inf A, where $A \subset R$

(b) Prove using Theorem 3.1, an analogue of Theorem 3.1 for infremums.

6. Define monotone decreasing sequence of real numbers and prove an analogue of Theorem 3.2 concerning such sequences.

7. (a) Show that $L \times Q$ is a normed linear space under the definitions given.

(b) Show that $\| X \| \leq \| (X, Y) \|$; $\| Y \| \leq \| (X, Y) \|$.

CHAPTER 3

Functions on Banach Spaces

1. Continuous Functions

In this section L and Q are two normed linear spaces over the same field F.

Definition 1.1. Let f be a function from $A \subset L$ into Q. Then f is *continuous* on A if:

$$X_n \in A \text{ and } X_n \to X \Rightarrow f(X_n) \to f(X)$$

Definition 1.2. Let f be a function defined on A, then the *image* of A under f, $f[A]$, is:

$$f[A] = \{Y \mid Y = f(X) \text{ and } X \in A\}$$

Theorem 1.1. If f is continuous on a compact set A then $f[A]$ is compact.

Proof: Let $Y_n \in f[A]$ be a sequence. We must show that $\{Y_n\}$ has a convergent subsequence. For each n, then exists $X_n \in A$ such that

$$Y_n = f(X_n)$$

hence for a suitable subsequence

$$X_{n_k} \to X \in A$$

Therefore

$$Y_{n_k} = f(X_{n_k}) \to f(X) \in f[A]$$

Corollary 1.2. Let f be continuous on a compact set A. Then there is a number $M > 0$ such that

$$X \in A \Rightarrow \|f(X)\| \leq M$$

Proof: $f[A]$ is compact $\Rightarrow f[A]$ is bounded.

Corollary 1.3. Let Q be R, and let f be continuous on a compact set A. Then f takes on a maximum and a minimum value on A.

Proof: $f[A]$ is compact $\Rightarrow f[A]$ is bounded and closed. Let $M = \sup f[A]$. We must show that $M \in f[A]$. For each n, there is $Y_n \in f[A]$ such that

$$M \geq Y_n > M - 1/n$$

i.e.
$$|Y_n - M| < 1/n$$

Therefore, $Y_n \to M$; and $Y_n \in f[A]$

Therefore $M \in f[A]$ since $f[A]$ is closed, i.e. $M = f(X)$ for some $X \in A$. A similar proof holds for the minimum.

Definition 1.3. f is *uniformly continuous* on A means that for each $\epsilon > 0$ there is a $\delta > 0$ such that

$$X', X'' \in A \text{ and } \|X' - X''\| < \delta \Rightarrow \|f(X') - f(X'')\| < \epsilon$$

(Note that the value of δ is independent of the points X', X''.)

Theorem 1.4. If f is uniformly continuous on A, then f is continuous on A.

Proof: Let $X_n \in A$, $X_n \to X \in A$. We must show that $f(X_n) \to f(X)$.

Choose $\epsilon > 0$.

Then there is a $\delta > 0$ as in Definition 1.3.

Now, there exists on N such that

$$n > N \to \|X_n - X\| < \delta$$
So, $$n > N \to \|f(X_n) - f(X)\| < \epsilon$$
or $$f(X_n) \to f(X)$$

Corollary 1.5. If f is uniformly continuous on A, $X_n, X_n' \in A$, and $\|X_n - X_n'\| \to 0$, then $\|f(X_n) - f(X_n')\| \to 0$.

Proof: Choose $\epsilon > 0$. Then there exists a $\delta > 0$ such that Definition 1.3 holds. But, there exists an N such that

$$n > N \Rightarrow \|X_n - X_n'\| < \delta$$
Hence, $$n > N \Rightarrow \|f(X_n) - f(X_n')\| < \epsilon$$

Faulty Proof of Corollary 1.5 using only continuity: Let $X'_n \to X$.

Then, $X_n = X_n - X'_n + X'_n \to 0 + X = X$

Thus, $f(X_n) \to f(X)$

$$f(X'_n) \to f(X)$$

Hence, $f(X_n) - f(X'_n) \to 0$; the fault lies in the fallacious assumption that X exists such that $X'_n \to X$.

Example. Let $A = \{X \mid 0 < X \leq 1\}$ and let $f(X) = 1/X$. Then f is continuous on A. However it is not uniformly continuous since Corollary 1.5 is not satisfied.

For, *let* $X_n = \dfrac{1}{n^2}, \ X'_n = \dfrac{1}{n}$

Then $X_n - X'_n = \dfrac{1}{n^2} - \dfrac{1}{n} \to 0 - 0 = 0$

but $f(X_n) - f(X'_n) = n^2 - n = n(n-1) \to \infty$

Note. A is not a closed set and hence is not compact.

Theorem 1.6. Let f be continuous on a *compact* set A. Then f is uniformly continuous on A.

Proof: (by contradiction). Suppose f is continuous on A, but not uniformly continuous on A. Then there is an $\epsilon > 0$, such that for every $\delta > 0$, there are $X, X' \in A$ such that $\| X - X' \| < \delta$, but

$$\| f(X) - f(X') \| \geq \epsilon$$

Let ϵ_0 be such an ϵ. Setting $\delta = 1/n$, for every positive integer n, there are numbers $X_n, X'_n \in A$ such that $\| X_n - X'_n \| < 1/n$ and $\| f(X_n) - f(X'_n) \| \geq \epsilon_0$. From this we have $X_n - X'_n \to 0$. But, since A is compact, $\{X'_n\}$ has a subsequence $\{X'_{n_k}\}$ such that $X'_{n_k} \to X \in A$. Then, $X_{n_k} = (X_{n_k} - X'_{n_k}) + X'_{n_k} \to 0 + X = X$. Therefore, by the definition of continuity, $f(X'_{n_k}) \to f(X)$ and $f(X_{n_k}) \to f(X)$. Thus, $f(X_{n_k}) - f(X'_{n_k}) \to f(X) - f(X) = 0$. Hence, there surely is a number N ($= n_k$, for large enough k) such that $\| f(X_N) - f(X'_N) \| < \epsilon_0$. This is a contradiction, since $\| f(X_n) - f(X'_n) \| \geq \epsilon_0$ for all n.

2. The Spaces C^F (a, b)

If a, b are real numbers such that $a < b$, then we write:

$$[a, b] = \{X \in R \mid a \leqq X \leqq b\}$$

The set $[a, b]$ is called a *closed interval*.

Definition 2.1. $C^F(a, b) = \{f \mid f$ is a continuous function from $[a, b]$ into $F\}$. (Note that in the case $F = C$, the members of $C^F(a, b)$ may be regarded as parametric representation of continuous curves in the complex plane.)

In $C^F(a, b)$, we define:

(1) $(f+g)\,(X) = f(X)+g(X)$

(2) $(af)\,(X) = af(X)$, (where $a \in F$)

(3) $\|f\| = \sup\limits_{a \leqq X \leqq b} |f(X)|$.

Note that, by Corollary 1.3, $\|f\| = |f(X_0)|$ for some $X_0 \in [a, b]$.

With the above definitions, it is easy to see that the spaces $C^F(a, b)$ are normed linear spaces, e.g.:

$$
\begin{aligned}
\|f+g\| &= \sup_{a \leqq X \leqq b} |(f+g)X| \\
&= \sup_{a \leqq X \leqq b} |f(X)+g(X)| \\
&\leqq \sup_{a \leqq X \leqq b} [\,|f(X)| + |g(X)|\,] \\
&\leqq \sup_{a \leqq X \leqq b} |f(X)| + \sup_{a \leqq X \leqq b} |g(X)| \\
&= \|f\| + \|g\|
\end{aligned}
$$

Also:

$$
\begin{aligned}
\|af\| &= \sup_{a \leqq X \leqq b} |(af)(X)| \\
&= \sup_{a \leqq X \leqq b} |af(X)| \\
&= \sup_{a \leqq X \leqq b} |a| \cdot |f(X)| \\
&= |a| \sup_{a \leqq X \leqq b} |f(X)| \\
&= |a| \cdot \|f\|
\end{aligned}
$$

The rest of the verification is left to the student.

Theorem 2.1. $\|f_n - f\| \to 0 \Leftrightarrow f_n(X) \to f(X)$ uniformly in $[a, b]$.

Proof $\Rightarrow$: $\|f_n - f\| = \sup\limits_{a \leq X \leq b} |f_n(X) - f(X)|$. Choose $\epsilon > 0$. Then there is N such that

$$n > N \Rightarrow \|f_n - f\| < \epsilon$$

Then, for each $X \in [a, b]$,

$$n > N \Rightarrow |f_n(X) - f(X)| \leq \sup\limits_{a \leq X \leq b} |f_n(X) - f(X)| = \|f_n - f\| < \epsilon$$

Proof $\Leftarrow$: Choose $\epsilon > 0$, then, by the definition of uniform convergence there is an N such that $X \in [a, b]$ and

$$n > N \Rightarrow |f_n(X) - f(X)| < \epsilon$$

For each n, there exists a number, $X^{(n)} \in [a, b]$ such that

$$\sup\limits_{X \in [a, b]} |f_n(X) - f(X)| = |f_n(X^n) - f(X^n)|$$

Hence, $n > N \Rightarrow \|f_n - f\| < \epsilon$.

Theorem 2.2. $C^F(a, b)$ is a Banach space.

Proof: Let $\{f_n\}$ be a Cauchy sequence. Then

$$\lim\limits_{\substack{n \to \infty \\ m \to \infty}} \|f_n - f_m\| = 0$$

Now, for each $X \in [a, b]$, $|f_n(X) - f_m(X)| \leq \|f_n - f_m\| \to 0$. Therefore, for each $X \in [a, b]$ the sequence $\{f_n(X)\}$ of elements of F is a Cauchy sequence. For each $X \in [a, b]$, there is a number $f(X) \in F$ such that $f_n(X) \to f(X)$. But, for each $\epsilon > 0$, there is an N such that

$$m, n > N \Rightarrow |f_n(X) - f_m(X)| < \epsilon$$

Letting $m \to \infty$: $n > N \Rightarrow |f_n(X) - f(X)| \leq \epsilon$. Therefore $\{f_n\}$ converges uniformly to f. Therefore, by an elementary theorem, f is continuous. (Cf. e.g. Taylor's *Advanced Calculus*.) In other words, $f \in C^F(a, b)$. Finally, by Theorem 2.1 $\|f_n - f\| \to 0$. This proves that $C^F(a, b)$ is a Banach space.

3. Continuous Operators

Let L and Q be given normed linear spaces over F.

Definition 3.1. An *operator* is a function from L into Q. Note that an operator must be defined on the whole space.

Definition 3.2 A *functional* is an operator from L into F.

Definition 3.3. A operator T is called *linear* if:

(1) $T(X+Y) = TX+TY$.

(2) $T(aX) = aTX$.

Definition 3.4. An operator is *bounded* if there is an $M > 0$ such that $\| TX \| \leq M \| X \|$.

For linear operators, boundedness and continuity are equivalent as is shown in the following theorem:

Theorem 3.1. The linear operator T is continuous $\Leftrightarrow$ T is bounded.

Proof $\Leftarrow$:
$$\| TX-TX_n \| = \| T(X_n-X) \|$$
$$\leq M \| (X_n-X) \|$$

Therefore, $X_n \to X \Rightarrow TX_n \to TX$.

Proof $\Rightarrow$: Suppose T is continuous, but not bounded. Then, for each positive integer n, there is a point $X_n \in L$ such that

$$\| TX_n \| > n \| X_n \|$$

Let

$$Y_n = \frac{1}{n \| X_n \|} X_n$$

Then, $\| TY_n \| = \| \frac{1}{n \| X_n \|} T(X_n) \| = \frac{1}{n \| X_n \|} \| TX_n \| > \frac{n \| X_n \|}{n \| X_n \|} = 1$

I.e.
$$\| TY_n \| > 1$$

But,
$$\| Y_n \| = \frac{1}{\| n X_n \|} \cdot \| X_n \| = \frac{1}{n}$$

so that $Y_n \to 0$. By continuity, $TY_n \to 0$ so that $\| TY_n \|$ ultimately must be < 1. This is a contradiction.

Corollary 3.2. If T is linear and $Y_n \to 0 \Rightarrow TY_n \to 0$, then T is a continuous operator.

Proof: The proof just above that continuity of a linear operator implies boundedness uses only the fact that $Y_n \to 0 \Rightarrow TY_n \to 0$. Hence the proof applies to any T satisfying the hypothesis of the present Corollary. Hence, such a T must be bounded, and therefore, by Theorem 3.1, continuous.

4. Spaces of Bounded Linear Operators

The linear operators considered in this section will be bounded and hence continuous. (By Theorem 3.1 continuity and boundedness are equivalent for linear operators.)

Definition 4.1. $[L, Q] = \{T \mid T$ is a bounded linear operator from L into $Q\}$. For the sake of brevity, we will denote $[L, L]$ by $[L]$.

In this section it will be shown that, with suitable definitions of $+, \cdot,$ and norm, $[L, Q]$ forms a normed linear space if L and Q are normed linear spaces, and if in addition Q is a Banach space, then $[L, Q]$ forms a Banach space, whether or not L is one. It will further be shown that L a normed linear space implies $[L]$ a normed linear algebra, and L a Banach space implies $[L]$ a Banach algebra. In what follows L and Q are normed linear spaces unless otherwise stated.

We must first define a function $+$ from $([L, Q] \times [L, Q])$ into $[L, Q]$ and a function $\cdot$ from $(F \times [L, Q])$ into $[L, Q]$.

Definition 4.2. Let T, T' be operators. Then:
$T+T'$ is defined by $(T+T')X = TX+T'X$, and for $a \in F$, aT is defined by $(aT)X = a(TX)$.

Theorem 4.1. If T, T' are linear, so are $T+T'$ and aT.

Proof: We must show that the two requirements of Definition 3.3 are satisfied. First for $T + T'$:

$$
\begin{aligned}
(1)\ (T+T')(X+Y) &= T(X+Y)+T'(X+Y) && \text{(from Def. 4.2)} \\
&= (TX+TY)+(T'X+T'Y) && \text{(since each operator} \\
& && \text{is linear)} \\
&= (TX+T'X)+(TY+T'Y) && \text{(Q is a linear space)} \\
&= (T+T')X+(T+T')Y && \text{(from Def. 4.2)}
\end{aligned}
$$

(2) $(T+T')(bX)$ $= T(bX)+T'(bX)$ (Def. 4.2)

                 $= bTX+bT'X$ (each operator

                                            linear)

                 $= b(TX+T'X)$ (Q a linear space)

                 $= b(T+T')X$ (Def. 4.2)

Next for aT:

(1) $(aT)(X+Y)$ $= a(T(X+Y))$ (Def. 4.2)

                 $= a(TX+TY)$ (T linear)

                 $= a(TX)+a(TY)$ (Q a linear space)

                 $= (aT)X+(aT)Y$ (Def. 4.2)

(2) $(aT)(bX)$ $= a(T(bX)$ (Def. 4.2)

              $= ab(TX)$ (T linear)

              $= ba(TX)$ (F abelian)

              $= b(aT)X$ (Def. 4.2)

Theorem 4.2. $T, T' \in [L, Q] \Rightarrow T+T' \in [L, Q]$ and $aT \in [L, Q]$.

Proof: To be in $[L, Q]$ means to be linear and bounded. Theorem 4.1 shows that these elements are linear; we must now show them to be bounded.

$$\| (T+T')(X) \| = \| TX+T'X \| \leq \| TX \| + \| T'X \|$$

$$\leq M \| X \| + M' \| X \| = (M+M') \| X \|.$$

$$\| (aT)(X) \| = \| aTX \| = | a | \cdot \| TX \|$$

$$\leq | a | M \| X \|.$$

Since there is closure under $+$ and $\cdot$, $[L, Q]$ is itself a linear space. We now define norm in L, Q.

Definition 4.3. $\| T \| = \sup_{\| X \| = 1} \| TX \|$ for $T \in [L, Q]$.

Corollary 4.3. $\| T \| < \infty$.

Proof: $\| TX \| \leq M \| X \|$ (T is bounded).

$$\| X \| = 1 \Rightarrow \| TX \| \leq M.$$

$$\| T \| \leq M < \infty.$$

Corollary 4.4. $\|T\| = \sup_{\|X\|\leq 1} \|TX\|$.

Proof: Let $K = \sup_{\|X\|\leq 1} \|TX\|$.

$\|T\| \leq K$ ($\|T\|$ is defined as the supremum of a subset of the set of which K is the supremum.)

Let $\qquad\qquad\qquad \|X_0\| \leq 1$.

Let $\qquad\qquad\qquad X = \dfrac{1}{\|X_0\|} X_0$

$$\|X\| = 1$$
$$X_0 = \|X_0\| X$$
$$\|TX_0\| = \|T(\|X_0\| X)\|$$
$$= \|X_0\| \|TX\|$$
$$\leq 1 \cdot \|T\|.$$

Thus, $\qquad\qquad\qquad K \leq \|T\|$.

Hence, $\qquad\qquad\qquad K = \|T\|$.

Theorem 4.5. $\|TX\| \leq \|T\| \cdot \|X\|$.

Proof: For $X = 0$, the result is clear, since
$$T(0) = T(0+0) = T(0)+T(0)$$
$$T(0) = 0$$

Otherwise, let $\qquad X_0 = \dfrac{1}{\|X\|} X$
$$\|X_0\| = 1$$
$$\|TX_0\| \leq \|T\|$$
$$\|TX_0\| = \left\|T\left(\dfrac{1}{\|X\|}\right)X\right\| = \left\|\dfrac{1}{\|X\|}TX\right\|$$
$$= \dfrac{1}{\|X\|}\|TX\|$$
$$\|TX\| \leq \|X\| \cdot \|T\|$$

Corollary 4.6. $\|T\| = \sup\limits_{X \neq 0} \dfrac{\|TX\|}{\|X\|}$

Proof: Let $K = \sup\limits_{X \neq 0} \dfrac{\|TX\|}{\|X\|}$

Taking $\|X\| = 1$, $\|T\| \leq K$

Letting $X_0 \neq 0$, we have

$$\|TX_0\| \leq \|T\| \cdot \|X_0\| \quad \text{(Theorem 4.5)}$$

$$\frac{\|TX_0\|}{\|X_0\|} \leq \|T\|$$

$$K \leq \|T\|$$

Therefore, $\|T\| = K$.

Theorem 4.7. $[L, Q]$ is a normed linear space.

Proof: $[L, Q]$ has already been shown to be a linear space. Now we must show that the four conditions of Definition $2-1.2$ are satisfied.

(1) $\|T\| = \sup\limits_{\|X\|=1} \|TX\| \geq 0$

(2) $\|T + T'\| = \sup\limits_{\|X\|=1} \|(T + T')X\|$

$$\leq \sup\limits_{\|X\|=1} (\|TX\| + \|T'X\|)$$

$$\leq \sup\limits_{\|X'\|=1} \|TX'\| + \sup\limits_{\|X''\|=1} \|T'X''\|$$

$$= \|T\| + \|T'\|$$

(3) $\|aT\| = \sup\limits_{\|X\|=1} \|(aT)X\| = \sup\limits_{\|X\|=1} |a| \cdot \|TX\|$

$$= |a| \sup\limits_{\|X\|=1} \|TX\|$$

$$= |a| \cdot \|T\|$$

(4) $\|T\| = 0 \Leftrightarrow \sup\limits_{X \neq 0} \dfrac{\|TX\|}{\|X\|} = 0 \Leftrightarrow \dfrac{\|TX\|}{\|X\|} = 0$ for all $X \in L$

$$\Leftrightarrow \|TX\| = 0 \text{ for all } X \in L,$$

$$\Leftrightarrow TX = 0 \text{ for all } X \in L,$$

$$\Leftrightarrow T = 0$$

Theorem 4.8. If Q is a Banach space, so is $[L, Q]$.

Proof: Let $\{T_n\}$ be a Cauchy sequence

$$\lim_{\substack{m \to \infty \\ n \to \infty}} \| T_m - T_n \| = 0$$

Consider $\{T_n X\}$, where $X \in L$ and $T_n X \in Q$.

$\| T_m X - T_n X \| = \| (T_m - T_n)X \| \leq \| T_m - T_n \| \cdot \| X \| \to 0$ as $m, n \to \infty$
$\{T_n X\}$ is therefore a Cauchy sequence.

Let, $\qquad\qquad\qquad T_n X \to TX \quad$ (Q is a Banach space).

Next we inquire: is T in $[L, Q]$? To be in $[L, Q]$ it must be linear and bounded.

$$\begin{aligned}
(1) \quad T(X + Y) &= \lim_{n \to \infty} T_n(X + Y) = \lim_{n \to \infty} (T_n X + T_n Y) \\
&= \lim_{n \to \infty} T_n X + \lim_{n \to \infty} T_n Y \\
&= TX + TY.
\end{aligned}$$

$$\begin{aligned}
(2) \quad T(aX) &= \lim_{n \to \infty} T_n(aX) = \lim_{n \to \infty} a T_n X \\
&= a \lim_{n \to \infty} T_n X \\
&= aTX.
\end{aligned}$$

Therefore, T is linear.

$\{T_n\}$ is bounded (because it is a Cauchy sequence).
$\| T_n \| \leq M$
$\| T_n X \| \leq \| T_n \| \cdot \| X \| \leq M \| X \|$
$\| TX \| \leq M \| X \|$, so that T is bounded.

There remains now only to prove that T is the limit of $\{T_n\}$ as $n \to \infty$.

Choose $\epsilon > 0$. Then there is an N such that:

$$m, n > N \Rightarrow \| T_m - T_n \| < \epsilon$$
$$m, n > N \Rightarrow \| T_m X - T_n X \| = \| (T_m - T_n)X \|$$
$$\leq \| T_m - T_n \| \cdot \| X \| < \epsilon \| X \|$$

Letting, $n \to \infty$,

$$m > N \Rightarrow \| T_m X - TX \| \leq \epsilon \| X \|$$
$$m > N \Rightarrow \| (T_m - T) X \| \leq \epsilon \| X \|$$
$$m > N \Rightarrow \frac{\| (T_m - T) X \|}{\| X \|} \leq \epsilon$$
$$m > N \Rightarrow \| T_m - T \| \leq \epsilon \quad \text{(by Corollary 4.6)}$$

Therefore, $T_m \to T$, and $[L, Q]$ is a Banach space.

Definition 4.4. If $T, T' \in [L]$ then $(T \cdot T')X = T(T'X)$.

Corollary 4.9. $T, T' \in [L] \Rightarrow T \cdot T' \in [L]$.

Proof:
$$\begin{aligned}
(1) \quad (T \cdot T')(X + Y) &= T(T'(X + Y)) \\
&= T(T'X + T'Y) \\
&= T(T'X) + T(T'Y) \\
&= (T \cdot T')X + (T \cdot T')Y
\end{aligned}$$

$$\begin{aligned}
(2) \quad (T \cdot T')(aX) &= T(T'(aX)) \\
&= T(aT'X) \\
&= a(T \cdot T')(X)
\end{aligned}$$

$$\begin{aligned}
(3) \quad \| (T \cdot T')(X) \| &= \| T(T'X) \| \leq \| T \| \cdot \| T'X \| \\
&\leq \| T \| \cdot \| T' \| \cdot \| X \|
\end{aligned}$$

which proves that $T \cdot T'$ is bounded. But (1), (2), (3) imply that $T \cdot T' \in [L]$.

The proof of (3) just given proves:

Corollary 4.10. $\| TT' \| \leq \| T \| \cdot \| T' \|$.

Definition 4.5. The operator I on $[L]$ is defined by $IX = X$.

Corollary 4.11. I is a bounded linear operator.

Corollary 4.12. $IT = TI = T$ for $T \in [L]$. (The proofs are left for the student.)

Corollary 4.13. $\| I \| = 1$

Proof: $\| I \| = \sup_{\| X \| = 1} \| IX \| = \sup_{\| X \| = 1} \| X \| = 1$

Corollary 4.14. $[L]$ is a normed linear algebra, (the proof is left for the student).

Corollary 4.15. If L is a Banach space, $[L]$ is a Banach algebra.

Proof: $[L]$ is a normed linear algebra from Corollary 4.14 and a Banach space from Theorem 4.8.

Corollary 4.16. If $T_n \in [L]$ and $T \in [L]$ and $T_n \to T$ and $X \in L$ then $T_n X \to TX$.

Proof: $\| T_n X - TX \| = \| (T_n - T)X \| \leq \| T_n - T \| \cdot \| X \|$. But $T_n \to T \Rightarrow \| T_n - T \| \cdot \| X \| \to 0$. Therefore

$$\| T_n X - TX \| \to 0 \text{ implying } T_n X \to TX.$$

5. Existence Theorems for Integral Equations and Differential Equations

In this section we write $C(a, b)$ to mean $C^R(a, b)$.

Definition 5.1. An *acceptable Fredholm kernel function* is a continuous mapping, k, from $[a, b] \times [a, b]$ into R such that

$$M(b-a) < 1$$

where M is the maximum value taken on by $k(x, y)$ in $[a, b] \times [a, b]$. (Note that $[a, b] \times [a, b]$ is a square in the Cartesian plane. Since $k(x, y)$ is a continuous function on a compact set, it assumes a maximum value.) We now define K, the Fredholm operator associated with k. Given $X \in C(a, b)$, $KX = Y$ means:

$$Y(s) = \int_a^b k(s, t) X(t) dt$$

The question arises: Does $Y \in C(a, b)$? Let $s_n \to s$ then

$$Y(s_n) - Y(s) = \int_a^b [k(s_n, t) - k(s, t)] X(t) dt$$

Since k is continuous on a compact set, k is uniformly continuous on $[a, b] \times [a, b]$. Choose $\epsilon > 0$. Then there is a $\delta > 0$ such that if $\| (s_n, t) - (s, t) \| < \delta$ then

$$| k(s_n, t) - k(s, t) | < \frac{\epsilon}{2 \| X \| (b-a)}$$

(Note that if $\|X\| = 0$, $Y = 0$ and proof is trivial.)

$$\|(s_n, t) - (s, t)\| = \sqrt{(s_n - s)^2} = |s_n - s|$$

Therefore $|s_n - s| < \delta \Rightarrow \|k(s_n, t) - k(s, t)\| < \dfrac{\epsilon}{2\|X\|(b-a)}$

But $s_n \to s$. Therefore there is an N such that $n > N \Rightarrow |s_n - s| < \delta$. Therefore

$$n > N \Rightarrow |Y(s_n) - Y(s)| = \left| \int_a^b [k(s_n, t) - k(s, t)] X(t)\, dt \right|$$

$$\leqq \frac{\epsilon}{2\|X\|(b-a)} \|X\|(b-a) < \epsilon$$

This means that Y is a continuous function and hence $\in C(a, b)$.

Theorem 5.1. $K \in [C(a, b)]$. (That is, K is a bounded linear operator on $C(a, b)$.)

The proof is left for the student.

The *Fredholm integral equation* is

$$Y = (I - K)X$$

where I is the identity operator and K is the Fredholm operator.

Problem. Given $Y \in C(a, b)$ to find $X \in C(a, b)$ such that $Y = (I - K)X$. In integral form, the Fredholm equation is:

$$Y(s) = X(s) - \int_a^b k(s, t) X(t)\, dt$$

Corollary 5.2. $\|K\| < 1$.

Proof: $\|K\| = \sup_{\|X\| = 1} \|KX\|$

Let $Y = KX$ where $\|X\| = 1$.

Then,

$$Y(s) = \int_a^b k(s, t) X(t)\, dt$$

$$|Y(s)| \leqq M(b-a) \cdot 1$$

$$\|Y\| \leqq M(b-a), \quad \|KX\| \leqq M(b-a)$$

Hence $$\|K\| \leqq M(b-a) < 1$$

Corollary 5.3. The operator $I-K$ has an inverse given by the formula:

$$(I-K)^{-1} = I+K+K^2+\dots$$

Proof: $\|I-(I-K)\| = \|K\| < 1$

Now $[C(a, b)]$ (the algebra of all bounded linear operators on $C(a, b)$) is a Banach algebra, (by Corollary 4.15) so this result is an immediate consequence of Theorem 2-1.7.

Corollary 5.4. For each $Y \in C(a, b)$, there is a *unique* $X \in C(a, b)$ such that $Y = (I-K)X$, namely

$$X = \left(\sum_{n=0}^{\infty} K^n\right) \cdot Y$$

(Here K^0 is understood to mean I.)

Proof: Let

$$H = \sum_{n=0}^{\infty} K^n$$

By Corollary 5.3, $H(I-K) = (I-K)H = I$.

(1) *Existence.*
$$(I-K)(HY) = [(I-K) \cdot H]Y$$
$$= I \cdot Y = Y.$$

(2) *Uniqueness.* Let X satisfy:
$$(I-K)X = Y.$$

Then:
$$H[(I-K)X] = HY$$
$$[H(I-K)]X = HY$$
$$IX = HY$$
$$X = HY$$

Theorem 5.5. In Corollary 5.4, we may write

$$X = \sum_{n=0}^{\infty} (K^n Y)$$

Proof: Let

$$H_n = \sum_{k=0}^{n} K^k$$

Then, using Corollary 4.16

$$\left(\sum_{n=0}^{\infty} K^n\right) Y = \left(\lim_{n \to \infty} H_n\right) Y$$
$$= \lim_{n \to \infty} (H_n Y)$$
$$= \lim_{n \to \infty} \left(\sum_{k=0}^{n} K^k\right) Y$$
$$= \lim_{n \to \infty} \sum_{k=0}^{n} (K^k Y)$$
$$= \sum_{n=0}^{\infty} (K^n Y)$$

Definition 5.2

$$k_1(s, t) = k(s, t)$$
$$k_n(s, t) = \int_a^b k(s, u) k_{n-1}(u, t) \, du, \quad n > 1$$

Theorem 5.6.

$$Y = K^n X$$
$$\Leftrightarrow Y(s) = \int_a^b k_n(s, t) X(t) \, dt$$

Proof (by mathematical induction): Obvious for $n = 1$. Assume the result known for $n = q$. We must verify it for $n = q+1$.

$$Y = K^{q+1} X \Leftrightarrow Y = K(K^q X)$$

Hence using the induction hypothesis,

$$Y = K^{q+1} X \Leftrightarrow Y(s) = \int_a^b k(s, u) \left[\int_a^b k_q(u, t) X(t) \, dt \right] du$$
$$= \int_a^b \int_a^b k(s, u) k_q(u, t) X(t) \, dt \, du$$
$$= \int_a^b \int_a^b k(s, u) k_q(u, t) X(t) \, du \, dt$$
$$= \int_a^b X(t) \left[\int_a^b k(s, u) k_q(u, t) \, du \right] dt$$
$$= \int_a^b X(t) k_{q+1}(s, t) \, dt$$

Corollary 5.7. For the X of Corollary 5.4,

$$X(s) = \sum_{n=0}^{\infty} \int_a^b k_n(s, t) Y(t)\, dt$$

Corollary 5.8.

$$|k_n(s, t)| \le M^n (b-a)^{n-1}$$

Proof (by induction): Obvious for $n = 1$. Assume known for q.

$$
\begin{aligned}
|k_{q+1}(s, t)| &= \left| \int_a^b k(s, u)\, k_q(u, t)\, dt \right| \\
&\le (b-a) M \cdot M^q (b-a)^{q-1} \\
&= M^{q+1} (b-a)^q
\end{aligned}
$$

Corollary 5.9. For the X of Corollary 5.4,

$$X(s) = \int_a^b \sum_{n=0}^{\infty} k_n(s, t)\, Y(t)\, dt$$

Proof: $|k_n(s, t)| \le M^n (b-a)^{n-1}$

$\sum_{n=0}^{\infty} k_n(s, t)$ converges uniformly in $[a, b] \times [a, b]$ by the Weierstrass

M-test. Hence $\sum_{n=0}^{\infty} k_n(s, t) Y(t)$ converges uniformly in $[a, b] \times [a, b]$,
and term-by-term integration is legitimate.

We shall define the *Volterra operator* K so that for $X \in C(a, b)$

$$Y = KX \Leftrightarrow Y(s) = \int_a^s k(s, t) X(t)\, dt$$

The conditions on k will now be weaker than in the case of the Fredholm operator.

Definition 5.3. k is an *acceptable Volterra kernel* if it is a continuous real valued function on the triangular region:

$$\{(s, t) \mid a \le t \le s \le b\}$$

Definition 5.4. $k_1(s, t) = k(s, t)$.

$$k_n(s, t) = \int_t^s k(s, u)k_{n-1}(u, t)\, du \text{ for } n > 1, \text{ where } a \leqq t \leqq s \leqq b$$

Theorem 5.10.

$$Y = K^n X \Leftrightarrow Y(s) = \int_a^s k_n(s, t)\, X(t)\, dt$$

Proof: By mathematical induction:

(*a*) for $n = 1$, the result is obvious from the definition of K.

(*b*) assume the result is true for $n = q$ then

$$Y = K^{q+1}X = K(K^q X)$$

$$\Leftrightarrow Y(s) = \int_a^s k(s, u)\left[\int_a^u k_q(u, t)\, X(t)\, dt\right] du$$

from the definition of K and the induction hypothesis. Hence,

$$Y(s) = \int_a^s \int_a^u k(s, u)\, k_q(u, t)\, X(t)\, dt\, du$$

Now the order of integration can be changed if the limits are selected properly. If this is done, (see figure below)

$$Y(s) = \int_a^s \int_t^s k(s, u)\, k_q(u, t)\, X(t)\, du\, dt$$

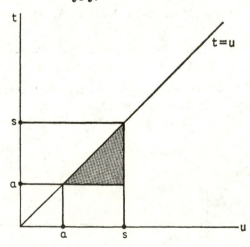

Hence,
$$Y(s) = \int_a^s X(t)\left[\int_t^s k(s,u)\,k_q(u,t)\,du\right]dt$$

or
$$Y(s) = \int_a^s k_{q+1}(s,t)\,X(t)\,dt$$

by the definition of $k_n(s,t)$.

Theorem 5.11. Let $M = \sup |k(s,t)|$ where the supremum is taken over all (s,t) on which k is defined, i.e. all (s,t) such that $a \leqq t \leqq s \leqq b$. Then
$$|k_n(s,t)| \leqq \frac{M^n(s-t)^{n-1}}{(n-1)!}$$

Proof: By mathematical induction:

(a) For $n = 1$, the result is obvious from the definition of M and the definition of supremum.

(b) Assume the result is true for $n = q$. Then
$$|k_{q+1}(s,t)| \leqq \left|\int_t^s k(s,u)\,k_q(u,t)\,du\right|$$

by the definition of k_n.
Therefore
$$|k_{q+1}(s,t)| \leqq \int_t^s |k(s,u)|\cdot|k_q(u,t)|\,du \leqq \frac{MM^q}{(q-1)!}\int_t^s (u-t)^{q-1}\,du$$

by the induction hypothesis. Hence
$$|k_{q+1}(s,t)| \leqq \frac{M^{q+1}}{(q-1)!}\frac{(u-t)^q}{q}\bigg]_t^s = \frac{M^{q+1}}{q!}(s-t)^q$$

Corollary 5.12. $\quad |k_n(s,t)| \leqq \dfrac{M^n(b-a)^{n-1}}{(n-1)!}$

Proof: Theorem 5.11 and the fact that $(s-t) \leqq (b-a)$.

Theorem 5.13. $\quad \|K^n\| \leqq \dfrac{M^n(b-a)^n}{n!}$

Proof: $K^n = \sup_{\|X\|=1} \|K^nX\|$

Now let $Y = K^n X$ for $\| X \| = 1$.

Then
$$Y(s) = \int_a^s k_n(s, t) X(t)\, dt \quad \text{by Theorem 5.10.}$$

Hence,
$$|Y(s)| \leq \int_a^s \frac{M^n(s-t)^{n-1}}{(n-1)!}(1)\, dt$$

by Theorem 5.11 and since $\| X \| = 1$.

$$|Y(s)| \leq \frac{M^n}{(n-1)!}\left[\frac{-(s-t)^n}{n}\right]_a^s = \frac{M^n(s-a)^n}{n!} \leq \frac{M^n(b-a)^n}{n!}$$

Hence
$$\| K^n X \| \leq \frac{M^n(b-a)^n}{n!}$$

for all X such that $\| X \| = 1$.

Therefore,
$$\| K^n \| \leq \frac{M^n(b-a)^n}{n!}$$

from the definition of $\| K^n \|$.

Theorem 5.14. The series $\sum_{n=0}^{\infty} K^n$ converges to an operator H such that $H(I-K) = (I-K)H = I$.

Proof: By Theorem 5.13 and the comparison test

$$\left(\text{since } \sum_{n=0}^{\infty} \frac{M^n(b-a)^n}{n!} \text{ converges to } e^{M(b-a)}\right),$$

$\sum_{n=0}^{\infty} \| K^n \|$ converges. Hence, $\sum_{n=0}^{\infty} K^n$ converges, since $K \in [C(a, b)]$ which is a Banach space and from Corollary 2-1.4, in a Banach space, if $\sum_{n=0}^{\infty} \| X_n \|$ converges, then $\sum_{n=0}^{\infty} X_n$ converges.

The rest of the proof follows as in Theorem 2-1.7:

$$H = \sum_{n=0}^{\infty} K^n = I + \sum_{n=1}^{\infty} K^n$$

Hence, $-HK = -\lim_{m \to \infty} \sum_{n=0}^{m} K^n \cdot K = -\lim_{m \to \infty} \sum_{n=0}^{m} K^{n+1} = -\sum_{n=1}^{\infty} K^n$

Also $-KH = -\lim_{m \to \infty} \sum_{n=0}^{m} K \cdot K^n = -\lim_{m \to \infty} \sum_{n=0}^{m} K^{n-1} = -\sum_{n=1}^{\infty} K^n$

Thus $\quad -HK = -KH\quad$ and $\quad H(I-K) = (I-K)H = H - \sum_{n=1}^{\infty} K^n$

Therefore, $H(I-K) = (I-K)H = I$.

Theorem 5.15. (Existence and uniqueness theorem for the Volterra equation.) For each $Y \in C(a, b)$ there is exactly one $X \in C(a, b)$ such that $Y = (I-K)X$. In fact,

$$X = HY = \left(\sum_{n=o}^{\infty} K^n\right)Y = \sum_{n=o}^{\infty} K^n Y$$

Or $\quad X(s) = \sum_{n=0}^{\infty} \int_a^s k_n(s,t) Y(t)\, dt = \int_a^s \sum_{n=0}^{\infty} k_n(s,t) Y(t)\, dt$

Proof: Exactly as for the Fredholm equation. See Corollary 5.4, Theorem 5.5, Corollary 5.7 and Corollary 5.9.

Theorem 5.16. (Existence and uniqueness theorem for second order linear differential equations with initial conditions.) Let $X, A_1, A_2 \in C$ (a, b) and let m and n be any real numbers. Then there exists exactly one function $Y \in C(a, b)$ such that:

(1) $Y''(s) + A_1(s) Y'(s) + A_2(s) Y(s) = X(s)$,

(2) $Y(a) = m$,

(3) $Y'(a) = n$.

Proof: Assume that we have a Y satisfying these conditions and define $Z(s) = Y''(s)$.

Then $\quad \int_a^s Z(t)\, dt = Y'(s) - Y'(a) = Y'(s) - n$

Also, $\quad \int_a^s \int_a^u Z(t)\, dt\, du = \int_a^s [Y'(u) - n]\, du = Y(s) - Y(a) - n(s-a)$

Hence $\quad \int_a^s \int_a^u Z(t)\, dt\, du = Y(s) - m - n(s-a)$

However, by changing the order of integration

$$\int_a^s \int_a^u Z(t)\, dt\, du = \int_a^s \int_t^s Z(t)\, du\, dt = \int_a^s Z(t)(s-t)\, dt$$

Therefore, $\quad Y(s) - m - n(s-a) = \int_a^s Z(t)(s-t)\, dt$

Substituting in the differential equation,

$$Z(s) + A_1(s)\left[n + \int_a^s Z(t)\,dt\right]$$

$$+ A_2(s)\left[m + n(s-a) + \int_a^s Z(t)(s-t)\,dt\right] = X(s)$$

This can be written,

$$Z(s) - \int_a^s \left[-A_1(s) - A_2(s)(s-t)\right]Z(t)\,dt$$

$$= X(s) - nA_1(s) - A_2(s)\left[m + n(s-a)\right]$$

But this is a form of the Volterra integral equation

$$Z(s) - \int_a^s k(s,t)\,Z(t)\,dt = V(s)$$

which has a unique solution as given in Theorem 5.15.

Moreover, as is easily seen, if $Z(s)$ satisfies this equation, then

$$Y(s) = \int_a^s\left[\int_a^v Z(u)\,du + n\right]dv + m$$

satisfies (1), (2), and (3).

6. The Hahn-Banach Extension Theorem

In this section L is a fixed normed linear space over F, where F can be either R or C.

Suppose that we have $V_1 \subset V_2 \subset L$ where V_1 and V_2 are linear subspaces of L. Let f_1 be a function defined on V_1 and f_2 a function defined on V_2. Then $f_1 \subset f_2$ means

$$(X, Y) \in f_1 \Rightarrow (X, Y) \in f_2$$

or

$$f_1(X) = Y \Rightarrow f_2(X) = Y$$

That is, f_2 is an extension of f_1 if for each X for which $f_1(X)$ is defined, $f_2(X)$ is also defined and $f_2(X) = f_1(X)$.

Theorem 6.1. Let $V_1 \subset V_2 \subset L$, where V_1 and V_2 are linear subspaces of L. Let f_1 and f_2 be bounded linear functionals (cf. Definition 3.2) on V_1 and V_2 respectively, and let $f_1 \subset f_2$. Then $\|f_1\| \leq \|f_2\|$.

Proof: $\|f_1\| = \sup_{\substack{X \in V_1 \\ \|X\|=1}} |f_1(X)|$ (by definition)

$$= \sup_{\substack{X \in V_1 \\ \|X\|=1}} |f_2(X)| \quad \text{(because } f_1 = f_2 \text{ on } V_1\text{)}$$

$$\leq \sup_{\substack{X \in V_2 \\ \|X\|=1}} |f_2(X)| = \|f_2\|$$

Theorem 6.2. (Hahn-Banach extension theorem.) Let V be a linear subspace of L and let g be a bounded linear functional on V. Then, there is a bounded linear functional $G \supset g$, on L, such that $\|G\| = \|g\|$.

Proof: (Case 1, $F = R$.)
Let $\mathscr{F}$ be the family of all f such that:

(1) f is a bounded linear functional on W which is a subspace of L.

(2) $g \subset f$.

(3) $\|f\| = \|g\|$.

Lemma. $\mathscr{F}$ is inductive.

Proof of Lemma: Let $\mathscr{C}$ be a chain, $\mathscr{C} \subset \mathscr{F}$. Let $k = \bigcup_{f \in \mathscr{C}} f$. We want to prove $k \in \mathscr{F}$. k is a function (that is, $(X, a) \in k$ and

$$(X, b) \in k \Rightarrow a = b)$$

because; $(X, a) \in f_1 \in \mathscr{C}$ and $(X, b) \in f_2 \in \mathscr{C}$, and since $\mathscr{C}$ is a chain one of these functions must be an extension of the other. Therefore, $a = b$.

Now, $k(X) = a \Leftrightarrow$ there is an $f \subset \mathscr{C}$ such that $f(X) = a$. For each $f \in \mathscr{C}$, let W_f be the space on which f is defined. Let $\mathscr{W} = \{W_f \mid f \in \mathscr{C}\}$. Let $W_k = \bigcup_{f \in \mathscr{C}} W_f = \bigcup_{W \in \mathscr{W}} W$. W_k is the set on which k is defined. W_k is a linear subspace of L. For, let $X, Y \in W_k$. Then $X \in W_{f_1}$, $Y \in W_{f_2}$. If, say, $W_{f_1} \subset W_{f_2}$, then $X, Y \in W_{f_2}$. Therefore, $aX + bY \in W_{f_2}$ and $aX + bY \in W_k$ so W_k is a linear subspace of L.

To complete the proof of the lemma we must show (1) k is a linear functional on W_k, (2) $g \subset k$, and (3) k is bounded and $\| k \| = \| g \|$.

$$X, Y \in W_k \Rightarrow X, Y \in W_{f_2}$$

$$\Rightarrow f_2(aX + bY) = af_2(X) + bf_2(Y)$$

$$\Rightarrow k(aX + bY) = ak(X) + bk(Y)$$

Thus k is a linear functional on W_k.

Let $g(X) = a$ for some $X \in V$. For each $f \in \mathscr{C}$, $f(X) = a$, that is $(X, a) \in f$. Thus, $(X, a) \in k$. Therefore, $k(X) = a$, or in other words $k \supset g$.

For each $X \in W_k$, there is an $f \in \mathscr{C}$ such that

$$\left| k(X) \right| = \left| f(X) \right| \leq \| f \| \cdot \| X \| = \| g \| \cdot \| X \|$$

Thus, k is bounded and $\| k \| \leq \| g \|$. But by Theorem 6.1, $\| k \| \geq \| g \|$ so that $\| k \| = \| g \|$.

So $k \in \mathscr{F}$ and this completes the proof of the lemma.

Returning to the proof of the theorem, we see by Zorn's lemma that $\mathscr{F}$ has some maximal element G. Since $G \in \mathscr{F}$, $\| G \| = \| g \|$ and $G \supset g$.

Let W_G be the space on which G is defined. We want to prove $W_G = L$.

Suppose there is a $Z_0 \in L$ and $Z_0 \notin W_G$.

Let $Q = \{ X + aZ_0 \mid a \in R$ and $X \in W_G \}$.

Then, Q is a linear subspace of L because

$$(a(X_1 + a_1 Z_0) + b(X_2 + a_2 Z_0)) = (aX_1 + bX_2) + (aa_1 + ba_2)Z_0 \in Q$$

Anytime $X_1 + a_1 Z_0 = X_2 + a_2 Z_0$ then $X_1 = X_2$ and $a_1 = a_2$ because

$$(a_1 - a_2)Z_0 = X_2 - X_1$$

Then, $$a_1 \neq a_2 \Rightarrow Z_0 = \frac{1}{a_1 - a_2}(X_2 - X_1) \in W_G$$

But, we are supposing that $Z_0 \notin W_G$.

Hence, $a_1 = a_2$ and $X_2 - X_1 = 0$ or $X_2 = X_1$.

We define H:

$$H(X + aZ_0) = H(X) + aH(Z_0) = G(X) + at$$

where t is some real number which will be determined. Obviously with any choice of t, H is a linear functional on Q. H will also be bounded if we can arrange matters so that

$$|H(X+aZ_0)| \leq \|g\| \cdot \|X+aZ_0\|$$

Consider the case when $a > 0$. Multiplying the inequality

$$H(X+aZ_0) \leq \|g\| \cdot \|X+aZ_0\|$$

by $1/a$ we get

$$\frac{1}{a}H(X+aZ_0) \leq \frac{1}{a}\|g\| \cdot \|X+aZ_0\|$$

$$\Leftrightarrow H\left(\frac{X}{a}+Z_0\right) \leq \|g\| \cdot \left\|\frac{X}{a}+Z_0\right\|$$

$$\Leftarrow H(u+Z_0) \leq \|g\| \cdot \|u+Z_0\| \text{ for } all\ u \in W_G$$

$$\Leftrightarrow G(u)+t \leq \|g\| \cdot \|u+Z_0\| \text{ for } all\ u \in W_G$$

$$\Leftrightarrow t \leq -G(u)+ \|g\| \cdot \|u+Z_0\| \text{ for } all\ u \in W_G$$

$$\Leftarrow t \leq \inf_{u \in W_G} \left[-G(u)+ \|g\| \cdot \|u+Z_0\|\right] = q$$

Next consider the case when $a < 0$. Multiplying this time by $-1/a$ we get

$$-\frac{1}{a}H(X+aZ_0) \leq -\frac{1}{a}\|g\| \cdot \|X+aZ_0\|$$

$$\Leftrightarrow H\left(-\frac{X}{a}-Z_0\right) \leq \|g\| \cdot \left\|-\frac{X}{a}-Z_0\right\|$$

$$\Leftarrow H(-u-Z_0) \leq \|g\| \cdot \|-u-Z_0\| \text{ for all } u \in W_G$$

$$\Leftrightarrow -G(u)-t \leq \|g\| \cdot \|u+Z_0\| \text{ for all } u \in W_G$$

$$\Leftrightarrow t \geq -G(u)- \|g\| \cdot \|u+Z_0\| \text{ for all } u \in W_G$$

$$\Leftarrow t \geq \sup_{u \in W_G} \left[-G(u)- \|g\| \cdot \|u+Z_0\|\right] = p$$

Note that q must be greater than p if our statements about t are to be satisfied. Thus it is necessary to show that for all u_1, u_2 in W_G

$$-G(u_1)- \|g\| \cdot \|u_1+Z_0\| \leq -G(u_2)+ \|g\| \cdot \|u_2+Z_0\|$$

$$\Leftrightarrow G(u_2)-G(u_1) \leq \|g\| \left[\|u_1+Z_0\|+\|u_2+Z_0\|\right]$$

The last statement is true because

$$G(u_2) - G(u_1) = G(u_2 - u_1)$$
$$\leq |G(u_2 - u_1)|$$
$$\leq \|g\| \cdot \|u_2 - u_1\|$$
$$= \|g\| \cdot \|(u_2 + Z_0) - (u_1 + Z_0)\|$$
$$\leq \|g\| \cdot [\|u_2 + Z_0\| + \|u_1 + Z_0\|]$$

Following the steps backward we see that for all a (including the case $a = 0$, for which it is obvious)

$$H(X + aZ_0) \leq \|g\| \cdot \|X + aZ_0\|$$

Now

$$|H(X + aZ_0)| = \pm H(X + aZ_0);$$

and,

$$-H(X + aZ_0) = H(-X - aZ_0)$$
$$\leq \|g\| \cdot \|-X - aZ_0\|$$
$$= \|g\| \cdot \|X + aZ_0\|$$

Thus

$$|H(X + aZ_0)| \leq \|g\| \cdot \|X + aZ_0\|$$

for all $X + aZ_0 \in Q$, i.e., H is bounded, and $\|H\| \leq \|g\|$. Hence, by Theorem 6.1, $\|H\| = \|g\|$. Moreover $H \supset G$ thus $H \in \mathscr{F}$.

Since G is maximal, $H = G$, and since $(Z_0, t) \in H$, it must also be true that $(Z_0, t) \in G$ which contradicts the assumption that $Z_0 \notin W_G$. Therefore $W_G = L$. This completes the proof for the case $F = R$.

Case 2. $F = C$ (Bohnenblust-Sobczyk)

Note that for each $X \in V$, $g(X)$ is a complex number. For the complex number $a + bi$, where $a, b \in R$, we write $a = \mathscr{R}(a + bi)$, $b = \mathscr{I}(a + bi)$.

Let

$$h(X) = \mathscr{R}(g(X)),$$
$$k(X) = \mathscr{I}(g(X))$$

V and L can be regarded as normed linear spaces over R. Now, $g(X) = h(X) + ik(X)$, so that
$$g(iX) = ig(X) = ih(X) - k(X)$$
$$g(iX) = h(iX) + ik(iX)$$

Thus $h(X) = k(iX)$ and $k(X) = -h(iX)$, so that we may write

$$g(X) = h(X) - ih(iX)$$

$$h(X + Y) = \mathcal{R}(g(X + Y))$$

$$= \mathcal{R}(g(X) + g(Y))$$

$$= \mathcal{R}(g(X)) + \mathcal{R}(g(Y))$$

$$= h(X) + h(Y)$$

Also, *for real a,*

$$h(aX) = \mathcal{R}(g(aX))$$

$$= \mathcal{R}(ag(X)) = a\mathcal{R}(g(X))$$

h is therefore a linear functional on V taken over $\mathcal{R}$.

Also $\qquad |h(X)| = |\mathcal{R}(g(X))| \le |g(X)| \le \|g\| \cdot \|X\|$

Thus h is bounded; $\|h\| \le \|g\|$.

By case 1, there is a bounded linear functional H on L over $\mathcal{R}$ such that $H \subset h$ and $\|H\| = \|h\|$.

Let $\qquad G(X) = H(X) - iH(iX)$

$$G(X + Y) = H(X + Y) - iH(i(X + Y))$$

$$= H(X) + H(Y) - i[H(iX + iY)]$$

$$= H(X) + H(Y) - i[H(iX) + H(iY)]$$

$$= H(X) - iH(iX) + H(Y) - iH(iY)$$

$$= G(X) + G(Y)$$

$$G((a + bi)X) = H((a + bi)X) - iH(i(a + bi)X)$$

$$= H(aX + biX) - iH(iaX - bX)$$

$$= H(aX) + H(biX) - iH(iaX) - iH(bX)$$

$$= aH(X) + bH(iX) - iaH(iX) + biH(X)$$

$$= (a + bi)H(X) - iH(iX)(a + bi)$$

$$= (a + bi)(H(X) - iH(iX))$$

$$= (a + bi)G(X)$$

Therefore G is a linear functional on L over C, and $G \supset g$.

Choose a fixed $X \in L$. Then, for suitable t,

$$G(X) = |G(X)| e^{it}$$

Let, $X_1 = e^{-it} X$

Then, $G(X_1) = e^{-it} G(X) = |G(X)|$

$$H(X_1) = \mathscr{R}(G(X_1)) = |G(X)|$$

Thus, $|G(X)| = H(X_1) = |H(X_1)| \leq \|H\| \cdot \|X_1\|$

$$= \|h\| \cdot \|X_1\|$$

$$\leq \|g\| \cdot \|X_1\| = \|g\| \cdot 1 \cdot \|X\|$$

$$= \|g\| \cdot \|X\|$$

Thus, $\|G\| \leq \|g\|$, so that by

Thoerem 6.1,

$$\|G\| = \|g\|.$$

Corollary 6.3. Let $X_0 \in L$. Then, there is a bounded linear functional G on L such that $G(X_0) = \|X_0\|$.

Proof. Let $V = \{aX_0 \mid a \in F\}$, $V \subset L$. The equations

$$a_1 X_0 + a_2 X_0 = (a_1 + a_2) X_0$$

$$b(aX_0) = (ba) X_0$$

show that V is a linear subspace of L.

Let $g(aX_0) = a \|X_0\|$. The equations

$$g(a_1 X_0 + a_2 X_0) = g((a_1 + a_2) X_0)$$

$$= (a_1 + a_2) \|X_0\|$$

$$= a_1 \|X_0\| + a_2 \|X_0\|$$

$$= g(a_1 X_0) + g(a_2 X_0)$$

and $g(b(aX_0)) = g((ba)X_0) = ba \|X_0\|$

$$= b(a \|X_0\|) = bg(aX_0)$$

show that g is a linear functional on V. The equalities

$$|g(aX_0)| = |a| \|X_0\| | = |a| | \|X_0\| = \|aX_0\|$$

show that g is bounded.

By the Hahn-Banach Extension Theorem then, there exists $G \supset g$, a bounded linear functional on L.

Corollary 6.4. Let $X_0 \in L$, $X_0 \neq 0$. Then, there is a bounded linear functional G on L such that $G(X_0) \neq 0$.

Proof: Use the G given by Corollary 6.3.

Corollary 6.5. If $X_0 \in L$ and $G(X_0) = 0$ for every bounded linear functional G on L then, $X_0 = 0$.

Corollary 6.6. Let $X_1, X_2 \in L$ and $X_1 \neq X_2$. Then, there is a bounded linear functional G on L such that $G(X_1) \neq G(X_2)$.

Proof. Let $X = X_2 - X_1$, $X \neq 0$. Use Corollary 6.4 to obtain a bounded linear functional G such that $G(X) \neq 0$. Then

$$G(X_2 - X_1) \neq 0. \quad G(X_2) - G(X_1) \neq 0. \quad G(X_2) \neq G(X_1).$$

7. The Existence of Green's Function

We now show how the Hahn–Banach extension theorem can be applied to a classical problem: the existence of Green's function. A bit more demand on the reader's knowledge of analysis will be made in this section than in the remainder of the book. This section can be entirely omitted without disturbing continuity.

As usual we write for the Laplacian operator

$$\nabla^2 = \frac{\partial^2}{\partial x^2} + \frac{\partial^2}{\partial y^2}$$

Definition 7.1. u is harmonic in the open set $D \subset R^2$ means $\nabla^2 u = 0$ on D, and $\partial^2 u/\partial x^2$, $\partial^2 u/\partial y^2$ are continuous in D.

Theorem 7.1. u is harmonic in $D \Leftrightarrow \exists v$ such that $u + iv$ is regular (analytic) in D.

Proof: See any book on complex variable theory.

Domain means open connected set. (Cf. Problem 14 below.) By the *boundary* of an open set D, we mean the set $\bar{D} - D$.

Theorem 7.2. (maximum modulus theorem). Let $f(z)$ be analytic in the bounded domain D and its boundary. Then, $|f(z)|$ takes on its maximum value on the boundary of D.

Proof: Cf. proof of Theorem 7.1.

Theorem 7.3. If u is harmonic on $\bar{D}$ where D is a domain, (i.e. in some open set $E \supset \bar{D}$) then u takes on its maximum and minimum values on the boundary of D.

Proof: It suffices to consider the maximum, since u is harmonic if and only if $-u$ is. Choose v such that $u + iv$ is analytic in D. Then, e^{u+iv} is analytic in D.

$|e^{u+iv}| = e^u |e^{iv}| = e^u$ takes on its maximum on the boundary; hence so does u.

Let D be a domain and let its boundary be the curve M; let $Q \in D$. Then, G_Q is a *Green's Function* on D if:

(1) $G_Q(P) = -\ln|P - Q| + k(P, Q)$ in D.

(2) $G_Q(P) = 0$ on M.

(3) G_Q is continuous on $\bar{D} - \{Q\}$ and harmonic on $D - \{Q\}$.

Let f be continuous on M, and let G_P be a Green's function on D. Furthermore, let

$$u(P) = \frac{1}{2\pi} \int_M f(Q) \frac{\partial G_P(Q)}{\partial n} \, ds$$

Then, $u = f$ on M, and u is harmonic in D: Cf. Nehari, *Conformal Mapping*. Thus, if the existence of a Green's function on D can be demonstrated, it will follow that Laplace's equation $\nabla^2 u = 0$ is solvable subject to arbitrary continuous boundary conditions.

It is easy to see that such boundary conditions determine a unique solution. For otherwise, if u_1, u_2 are two solutions, then $\nabla^2(u_2 - u_1) = 0$ in D and $u_2 - u_1 = 0$ on M. Hence, by Theorem 7.3, $u_2 - u_1 = 0$ in D.

We shall sketch a proof, due to Peter Lax (*Proc. Am. Math. Soc.*, 3 (1952), pp. 526–31) of the existence of a Green's function on any Cauchy domain D. (For the meaning of the term *Cauchy domain*, cf. Definition 5-3.2.)

Theorem 7.4. Let D be a Cauchy domain with boundary M. Then for each $Q \in D$, there is a Green's function G_Q on D.

Proof: Let B be the set of all continuous functions from M into R. Then, exactly as for $C^R[a, b]$, it is easy to see that under the definitions:

$$(f+g)(X) = f(X) + g(X)$$

$$(af)X = a \cdot f(X)$$

$$\|f\| = \sup_{X \in M} |f(X)|$$

B becomes a normed linear space and, in fact, a Banach space over R. Let B_0 be the set of all $f \in B$ for which $\exists u$ such that $u = f$ on M and u is harmonic in D. (E.g. If f is a constant $f \in B_0$.) By the above remarks on the uniqueness of solutions, for each $f \in B_0$, there is exactly one harmonic function u_f which extends f to D. Let us note that (since $\nabla^2(au + bv) = a\nabla^2 u + b\nabla^2 v$), B_0 is a linear subspace of B. Let Q be some fixed point in D. Let r_Q be defined for $f \in B_0$, by

$$r_Q(f) = u_f(Q)$$

Since, $u_{f+g} = u_f + u_g$, $u_{af} = au_f$, by uniqueness and by linearity of the Laplacian, r_Q is a linear functional on B_0. Moreover, by Theorem 7.3,

$$|r_Q(f)| = |u_f(Q)| \leq \sup_{z \in D} |u_f(z)|$$

$$\leq \sup_{z \in M} |f(z)|$$

$$= \|f\|$$

Hence r_Q is a bounded linear functional; moreover, $\|r_Q\| \leq 1$. But, $|r_Q(1)| = 1$. Hence, $\|r_Q\| = 1$. By the Hahn-Banach extension theorem (Theorem 6.2), there is a bounded linear functional $R_Q \supset r_Q$, defined on B, such that $\|R_Q\| = 1$.

Next, for each P in the plane, let g_P be defined by

$$g_P(z) = \ln |z - P|$$

for $z \in M$. If $P \notin \bar{D}$, then $g_P \in B_0$, since it has the harmonic extension to D, u_{g_P} where

$$u_{g_P}(Q) = \ln |Q - P|$$

Thus, $$r_Q(g_P) = \ln |Q - P|$$

Whether or not $P \in \bar{D}$, so long as $P \notin M$, g_P is continuous on M, so that

$$P \notin M \Rightarrow g_P \in B$$

Hence, for $P \notin M$, we may define k_Q by

$$k_Q(P) = R_Q(g_P)$$

For $P \in M$, we define

$$k_Q(P) = \ln |Q - P|$$

Now, the Green's function G_Q may be defined by

$$G_Q(P) = -\ln |P - Q| + k_Q(P)$$

Then, G_Q clearly satisfies conditions (1) and (2) in the defining characterization of the Green's function. In order to verify condition (3), we shall require the lemmas:

Lemma 1. The operators R_Q and ∇^2 commute: $\nabla^2 R_Q = R_Q \nabla^2$.

Lemma 2. Let $z \in M$, and $P_0 \in M$. For each $P \in D$, let P' be the mirror image of P in the tangent line to M at the point of M nearest to P. Then,

$$\lim_{P \to P_0} \frac{|z - P'|}{|z - P|} = 1$$

uniformly in z.

Assume for the moment, the validity of these lemmas; we may proceed as follows:

In D,

$$\begin{aligned}
\nabla^2 k_Q(P) &= \nabla^2 R_Q(g_P) \\
&= R_Q \nabla^2 g_P \\
&= R_Q 0 \\
&= 0
\end{aligned}$$

Hence, G_Q is harmonic in $D - \{Q\}$.

We shall show that G_Q is continuous on $\bar{D} - \{Q\}$ by showing that it is continuous across M. By Lemma 2, if P and P' are related as in the hypotheses of Lemma 2,

$$\sup_{z \in M} \left| \ln \frac{|z - P|}{|z - P'|} \right| \to 0 \quad \text{as} \quad P \to P_0$$

That is $\| g_P - g_{P'} \| \to 0$ as $P \to P_0$.
Hence,

$$
\begin{aligned}
\left| k_Q(P) - k_Q(P') \right| &= \left| R_Q(g_P) - R_Q(g_{P'}) \right| \\
&= \left| R_Q(g_P - g_{P'}) \right| \\
&\leq \| R_Q \| \cdot \| g_P - g_{P'} \| \\
&= \| g_P - g_{P'} \| \to 0
\end{aligned}
$$

as $P \to P_0$.

It remains only to prove Lemmas 1 and 2.

Lemma 1 is an immediate consequence of:

Lemma 3. Let f be a continuous function on $M \times C$. Let L be a bounded linear functional on B. For $z \in C$ let $z = x + iy$. For each $z \in C$, let $\partial f / \partial x$ be continuous on M. Then $\partial / \partial x (Lf)$ exists and equals $L(\partial f / \partial x)$.

Proof of Lemma 3. Let $g(x, y) = Lf(P, x + iy)$. Then

$$
\begin{aligned}
\frac{g(x, y_0) - g(x_0, y_0)}{x - x_0} &= \frac{1}{x - x_0} \left[Lf(P, x + iy_0) - Lf(P, x_0 + iy_0) \right] \\
&= \frac{1}{x - x_0} L \left[f(P, x + iy_0) - f(P, x_0 + iy_0) \right] \\
&= L \left[\frac{f(P, x + iy_0) - f(P, x_0 + iy_0)}{x - x_0} \right]
\end{aligned}
$$

Hence,

$$\frac{g(x, y_0) - g(x_0, y_0)}{x - x_0} - Lf_x(P, x_0 + iy_0)$$

$$= L \left[\frac{f(P, x + iy_0) - f(P, x_0 + iy_0)}{x - x_0} - f_x(P, x_0 + iy_0) \right]$$

$$\to L(0) = 0 \text{ because } L \text{ is continuous.}$$

Proof of Lemma 2: Given $\epsilon > 0$. We must determine a number $\delta > 0$ such that $\left| P - P_0 \right| < \delta \Rightarrow$

$$\left| \left| \frac{z - P'}{z - P} \right| - 1 \right| < \epsilon$$

Since $\sqrt{x}$ is continuous at $x = 1$, there is a number q_1 such that

$$\left| x - 1 \right| < q_1 \quad \text{and} \quad x > 0 \Rightarrow \left| \sqrt{x} - 1 \right| < \epsilon$$

Since

$$\lim_{t \to 0} \frac{4 \sin t}{\cos^2 t} = 0$$

there is a number q_2 such that

$$\left| t \right| < q_2 \Rightarrow \frac{4 \left| \sin t \right|}{\cos^2 t} < q_1$$

Let M be given by $y = f(x)$ near P_0. Then $\operatorname{Arctan} f'(x)$ is continuous, hence uniformly continuous in a neighbourhood of P_0. Thus, there is an q_3 such that in this neighbourhood

$$\left| x_1 - x_2 \right| < q_3 \Rightarrow \left| \operatorname{Arctan} f'(x_1) - \operatorname{Arctan} f'(x_2) \right| < q_2$$

We distinguish two cases:

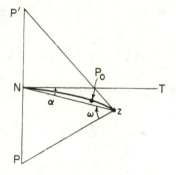

Case 1:

$$\left| z - P_0 \right| < q_3/2$$

Letting x_z be the abscissa of z, etc., we have:

$$\left| x_z - x_{P_0} \right| < q_3/2$$

Then, if P is sufficiently close to P_0 that $\left| x_N - x_{P_0} \right| < q_3/2$, then $\left| x_N - x_z \right| < q_3$. Now, the inclination of line NT to the x-axis is $\operatorname{Arctan} f'(x_N)$. By the mean value theorem, that of NZ is $\operatorname{Arctan} f'(x_i)$

where $|x_i - x_N| < |x_N - x_z| < q_3$. Hence, $|\alpha| < q_2$. Using the law of cosines and the fact that $|N - P'| = |N - P|$,

$$|z - P'|^2 - |z - P|^2 = 4|N - P| \cdot |N - z| \sin \alpha$$

$$\left| \left[\frac{|z - P'|}{|z - P|} \right]^2 - 1 \right| = 4 \frac{|N - P|}{|z - P|} \cdot \frac{|N - z|}{|z - P|} |\sin \alpha|$$

$$= 4 \frac{\sin \omega}{\cos \alpha} \cdot \frac{\cos (\alpha - \omega)}{\cos \alpha} |\sin \alpha|$$

$$\leq 4 \frac{|\sin \alpha|}{\cos^2 \alpha} < q_1$$

Hence,

$$\left| \frac{|z - P'|}{|z - P|} - 1 \right| < \epsilon$$

Thus, in this case, it suffices to take $\delta = q_3/2$.

Case 2: $|z - P_0| \geq q_3/2$

Then,

$$\left| \frac{|z - P'|}{|z - P|} - 1 \right| = \left| \frac{|z - P'| - |z - P|}{|z - P|} \right|$$

$$\leq \frac{|(z - P') - (z - P)|}{|z - P|}$$

$$= \frac{|P - P'|}{|(z - P_0) - (P - P_0)|}$$

$$\leq \frac{|P - P'|}{||z - P_0| - |P - P_0||}$$

$$\leq \frac{|P - P_0| + |P' - P_0|}{|(q_3/2) - |P' - P_0||}$$

But, as $P \to P_0$, $P' \to P_0$. Hence, this last expression (which is independent of z) $\to 0$.

Hence, $\exists\, q_4$ such that

$$|P - P_0| < q_4 \Rightarrow \left| \frac{|z - P'|}{|z - P|} - 1 \right| < \epsilon$$

Thus, to satisfy both cases take $\delta = \min (q_3/2, q_4)$.

Problems

1. Complete the proof that $C^F(a, b)$ is a normed linear space.

2. Prove Corollaries 4.11, 4.12, and 4.14.

3. Prove Theorem 5.1.

4. Using Corollary 5.4 solve the integral equation

$$\frac{5s}{6} = X(s) - \int_0^1 \frac{1}{2} st \cdot X(t)\, dt$$

Answer: $X(s) = s$.

5. Solve the integral equation,

$$s = X(s) - \int_0^s (t-s)\, X(t)\, dt$$

Answer: $X(s) = \sin s$.

6. For $X, Y \in R^n$, $X = (X_1, X_2, \ldots, X_n)$, $Y = (Y_1, Y_2, \ldots, Y_n)$ define the "dot product":

$$[X, Y] = X_1 Y_1 + X_2 Y_2 + \ldots + X_n Y_n$$

(Cf. Problem 3, Chapter 2.)

(a) For fixed a, let $f(X) = [a, X]$. Show that f is a bounded linear functional on R^n.

(b) Prove that for each bounded linear functional f on R^n, there is an $a \in R^n$ such that: $f(X) = [a, X]$.

Hint for b: What does f do to the elements

$$(1, 0, \ldots, 0), (0, 1, \ldots, 0), \ldots, (0, 0, \ldots, 1)?$$

7. (a) For fixed $M \in C^R(a, b)$, let

$$f(X) = \int_a^b X(t)\, M(t)\, dt$$

Show that f is a bounded linear functional on $C^R(a, b)$

(b) Are there any others?

8. (a) Let $B(a, b)$ consist of all real-valued functions f defined on $[a, b]$ for which

$$\sup_{a \leq t \leq b} |f(t)| < \infty$$

Show how to define the appropriate notions so as to make $B(a, b)$ a Banach space with subspace $C^R(a, b)$.

(b) Let

$$f(X)' = \int_a^b X(t)\, dt \quad \text{for} \quad X \in C_R(a, b)$$

Show that f is a bounded linear functional on $C^R(a, b)$. Evaluate $\|f\|$.

(c) Using (a), (b) and the Hahn-Banach theorem, what can we conclude?

9. Below are sets A in normed linear spaces as listed. In each case find $\bar{A}$.

	Space	A
(a)	R	$\{x \mid 0 < x < 1\}$
(b)	C	$\{z \mid 1 < |z| < 2\}$
(c)	$C^R(0, 1)$	$\left\{ X \left\| \int_0^1 X(t)\, dt \right\| < 1 \right\}$
(d)	$C^R(0, 1)$	$\{X \mid X(0) = 2\}$

10. Show that each of the following are bounded linear functionals on $C^R(0, 1)$ and find their norms:

(a) $f(X) = X(0)$,

(b) $f(X) = \int_0^1 X(t)\, e^{-t}\, dt$,

(c) $f(X) = 2X(0) + 3X(1)$.

11. Show that each of the following are bounded linear operators from R^2 into R^2 and find their norms:

(a) $T(a, b) = (b, a)$,

(b) $T(a, b) = ((a+b)/2, (a-b)/2)$,

(c) $T(a, b) = (6a, 6b)$.

12. Solve the integral equation:

$$u(s) = e^s - \tfrac{1}{2}\left(e - 1 - \int_0^1 u(t)\,dt\right)$$

13. (a) For which values of q do our results guarantee the existence of a solution for the integral equation:

$$X(s) - q \int_0^1 e^{s+t} X(t)\,dt = e^s$$

   (b) Solve the equation in (a) for the indicated range of values of q.

   (c) For which values of q is the solution obtained in (b) actually valid?

14. Let L, Q be normed linear spaces.

   Definition 1. $A \subset L$ is called *separated into B and C* if $B \subset L$, $C \subset L$, $B \neq \emptyset$, $C \neq \emptyset$, $A = B \cup C$, $\bar{B} \cap \bar{C} = \emptyset$.

   Definition 2. $A \subset L$ is called *connected* if it *can not* be separated into sets B and C.

   Theorem. Let $A \subset L$ be connected, and let f be a continuous function from A into Q. Then $f[A]$ is connected.

Problem. Prove the theorem.

CHAPTER 4

Homomorphisms on Normed Linear Spaces

I. Homomorphisms on Linear Spaces

In this section, L and Q are linear spaces over F.

Definition I.I. A *homomorphism* from L on to Q is a linear operator T on L such that for each $Y \in Q$, there is $X \in L$ such that $TX = Y$. In this case Q is called a *homomorphic image* of L.

Definition I.2. K_T, the *kernel* of a homomorphism T, is defined by

$$K_T = \{X \in L \mid TX = 0\}$$

Theorem I.I. K_T is a linear subspace of L.

Proof: We must show that those properties in the definition of a linear space which may be in doubt do hold.

(1) $0 \in K_T$.

(2) Suppose $X, Y \in K_T$.

$X, Y \in K_T$

$\Rightarrow TX = 0$ and $TY = 0$

$\Rightarrow TX + TY = 0$

$\Rightarrow T(X + Y) = 0$

$\Rightarrow X + Y \in K_T$.

(3) Suppose $X \in K_T$.

$X \in K_T \Rightarrow TX = 0 \Rightarrow aTX = 0 \Rightarrow T(aX) = 0 \Rightarrow aX \in K_T$.

Definition I.3. L and Q are *isomorphic* if there is a one-one homomorphism from L on to Q.

Problem: To find all (up to isomorphism) homomorphic images of L.

77

Definition 1.4. Let V be a subspace of L. Then, $X \equiv Y \bmod V$ means $X - Y \in V$. This is a relation on L

$$\{(X, Y) \mid X \equiv Y \bmod V\} \subset L \times L$$

Corollary 1.2. $X \equiv Y \bmod V$ is an equivalence relation on L.

Proof: (1) $X \equiv X \bmod V$

since $X - X = 0 \in V$. Hence the relation is reflexive.

(2) $X \equiv Y \bmod V \Rightarrow X - Y \in V \Rightarrow Y - X \in V \Rightarrow Y \equiv X \bmod V$. Hence it is symmetric.

(3) $X \equiv Y \bmod V$ and $Y \equiv Z \bmod V \Rightarrow X - Y \in V$ and

$$Y - Z \in V \Rightarrow X - Z = (X - Y) + (Y - Z) \in V \Rightarrow X \equiv Z \bmod V$$

Hence it is transitive.

Recalling the notation of section 3, Chapter 1:

$$[X] = \{Y \in L \mid X \equiv Y \bmod V\}$$

we have at once by Theorem 1–3.1:

Corollary 1.3. (1) $X \equiv Y \bmod V \Leftrightarrow [X] = [Y]$.

(2) $X \in [X]$.

(3) $[X] \cap [Y] \neq \emptyset \Rightarrow [X] = [Y]$.

Definition 1.5. $L/V = \{[X] \mid X \in L\}$, i.e. L/V is the set of all equivalence classes.

Corollary 1.4. (1) $X \equiv X' \bmod V$ and

$Y \equiv Y' \bmod V \Rightarrow X + Y \equiv X' + Y' \bmod V$.

(2) $X \equiv X' \bmod V \Rightarrow aX \equiv aX' \bmod V$.

Proof: (1) $(X + Y) - (X' + Y') = (X - X') + (Y - Y') \in V$.

(2) $aX - aX' = a(X - X') \in V$.

Definition 1.6. $[X] + [Y] = [X + Y]$ and $a[X] = [aX]$.

Corollary 1.5. L/V is a linear space with the definitions in Definition 1.6.

Proof: Left to the reader.

Theorem 1.6. The operator T defined by $TX = [X]$ is a homomorphism from L on to L/V with kernel V.

Proof: We must show that T is a linear operator.

$$T(X+Y) = [X+Y] = [X]+[Y] = TX+TY$$
$$T(aX) = [aX] = a[X] = aTX$$

Finally, $K_T = \{X \in L \mid TX = 0\} = \{X \in L \mid [X] = 0\}$
$$= \{X \in L \mid X \equiv 0 \bmod V\} = \{X \in L \mid X-0 \in V\}$$
$$= \{X \in L \mid X \in V\} = V.$$

Theorem 1.7. Let S be a homomorphism from L on to Q. Then Q is isomorphic to L/K_S.

Lemma 1. $[X] = [X'] \Rightarrow SX = SX'$.

Proof of Lemma: $[X] = [X'] \Rightarrow X \equiv X' \bmod K_S \Rightarrow X-X' \in K_S$
$\Rightarrow S(X-X') = 0 \Rightarrow SX - SX' = 0 \Rightarrow SX = SX'$.

Lemma 1 enables us to define: $R[X] = SX$.

Lemma 2. $R[X] = R[X'] \Rightarrow [X] = [X']$, i.e. R is one-one.

Proof of Lemma 2: $R[X] = R[X'] \Rightarrow SX = SX' \Rightarrow S(X-X') = 0$
$\Rightarrow X-X' \in K_S \Rightarrow [X] = [X']$.

To complete the proof of the theorem, we must finally show that sums and scalar products are preserved.

$$R([X]+[Y]) = R([X+Y]) = S(X+Y) = SX+SY = R[X]+R[Y]$$

Also $R(a[X]) = R([aX]) = S(aX) = aSX = aR[X]$.

Therefore Q is isomorphic to L/K_S.

2. Norms in a Quotient Space

In this section L is a normed linear space and V is a subspace.

Definition 2.1. $$\|[X]\| = \inf_{Z \in X} \|Z\|$$

Theorem 2.1. Let V be a *closed* subspace of L. Then Definition 2.1 makes L/V a normed linear space.

Proof: L/V is a linear space by Corollary 1.5. We must show that L/V is normed, i.e. must show:

(1) $\|[X]\| = \inf\limits_{Z \in [X]} \|Z\| \geqq 0$. Obvious, since $\|Z\| \geqq 0$.

(2) $\|[X]+[Y]\| \leqq \|[X]\| + \|[Y]\|$

By definition, $\|[X]+[Y]\| = \|[X+Y]\| = \inf\limits_{Z \in [X+Y]} \|Z\|$

But, $X_1 \in [X]$ and $Y_1 \in [Y] \Rightarrow X_1 + Y_1 \in [X+Y]$. Hence, we have

$$\|[X]+[Y]\| \leqq \inf\limits_{\substack{Y_1 \in [Y] \\ X_1 \in [X]}} \|X_1 + Y_1\| \leqq \inf\limits_{\substack{Y_1 \in [Y] \\ X_1 \in [X]}} [\|X_1\| + \|Y_1\|]$$

$$= \inf\limits_{X_1 \in [X]} \|X_1\| + \inf\limits_{Y_1 \in [Y]} \|Y_1\| = \|[X]\| + \|[Y]\|$$

(3) $\|[aX]\| = |a|\,\|[X]\|$. $Z \in [aX]$

$$\Rightarrow Z \equiv aX \bmod V \Rightarrow \frac{1}{a} Z \equiv X \bmod V \Rightarrow Z = a\left[\frac{1}{a}Z\right]$$

where $$\frac{1}{a}Z \in X$$

So $$\|[aX]\| = \inf\limits_{Z \in [aX]} \|Z\| = \inf\limits_{X_1 \in [X]} \|aX_1\|$$

$$= |a| \inf\limits_{X_1 \in [X]} \|X_1\| = |a|\,\|[X]\|$$

(4) $\qquad\qquad \|[X]\| = 0 \Leftrightarrow [X] = 0$. $\Leftarrow$ is obvious.

To verify the implication in the converse direction, we first note that $[X]$ is closed for each X. For, suppose that $Z_n \in [X]$, $Z_n \to Z$, and let $Y_n = Z_n - X$. Then, $Y_n \in V$, $Y_n \to Z - X$. Since V is closed, $Z - X \in V$. I.e. $Z \in [X]$. Thus, X is closed.

Now, suppose that $\|[X]\| = 0$. Then, $\inf\limits_{Z \in X} \|Z\| = 0$. Hence, for each positive integer n, there is a $Z_n \in [X]$ such that $\|Z_n\| < 1/n$. Thus, $Z_n \to 0$. Since X is closed, $0 \in X$, i.e. $[X] = [0]$.

We now consider the problem of determining under what conditions a homomorphism T is bounded (i.e. continuous). For when it is, the operations of analysis will be preserved under it.

Theorem 2.2. Let V be a closed subspace of L, and let T be defined by $TX = [X]$. Then, T is bounded (i.e. continuous).

Proof: We have to show that $\| TX \| \leq M \| X \|$ for some M.

$$\| TX \| = \| [X] \| = \inf_{Z \in [X]} \| Z \| \leq \| X \|.$$

(In fact, T has norm 1.)

Theorem 2.3. Let L, Q be normed linear spaces, and let T be a continuous homomorphism from L on to Q. Then K_T is closed.

Proof: Let $X_n \in K_T$, $X_n \to X$. Then $TX_n \to TX$ since T is continuous. But $TX_n = 0$. Therefore $TX = 0$, so that $X \in K_T$. Hence, K_T is closed.

Theorem 2.4. Let V be a closed subspace of the Banach Space L. Then, L/V is a Banach Space.

Proof. For this proof, we will adopt the notation

$$x, y, z \in L; \quad X, Y, Z \in L/V$$

The proof consists of showing that a Cauchy sequence in L/V converges.

Let $\{X_n\}$ be a Cauchy sequence. Then for each n, $\exists\ N_n$ such that $N_n < N_{n+1}$ and $r, p \geq N_n \Rightarrow \| X_p - X_r \| < 1/2^n$. In particular, $\| X_{N_{n+1}} - X_{N_n} \| < 1/2^n$. Now, let $Y_n = X_{N_n}$. Choose any $y_1 \in Y_1$. Choose some $y_2 \in Y_2$ such that $\| y_2 - y_1 \| < 2 \| Y_2 - Y_1 \|$. Choose some $y_3 \in Y_3$ such that $\| y_3 - y_2 \| < 2 \| Y_3 - Y_2 \|$ and in general, choose $y_{n+1} \in Y_{n+1}$ such that $\| y_{n+1} - y_n \| < 2 \| Y_{n+1} - Y_n \|$. Note that such a choice is always possible since otherwise, e.g., $2 \| Y_2 - Y_1 \|$ would be a lower bound for $\| z \|$, $z \in Y_2 - Y_1$ which is greater than $\| Y_2 - Y_1 \|$, the infremum, or greatest lower bound, of this class. The sequence of y's just described is a Cauchy sequence in L since $\| y_{n+p} - y_n \|$

$$= \| (y_{n+p} - y_{n+p-1}) + (y_{n+p-1} - y_{n+p-2}) + \ldots + (y_{n+1} - y_n) \|$$

$$\leq \sum_{k=1}^{p} \| y_{n+k} - y_{n+k-1} \|$$

$$\leq 2 \sum_{k=1}^{p} \| Y_{n+k} - Y_{n+k-1} \|$$

and remembering that we had $\| X_{N_{n+1}} - X_{N_n} \| < 1/2^n$, this becomes

$$< 2 \sum_{k=1}^{p} \frac{1}{2^{n+k-1}} = \frac{1}{2^{n-1}} \sum_{k=1}^{p} \frac{1}{2^{k-1}} < \frac{1}{2^{n-1}} \cdot 2 \to 0$$

Since $\{y_n\}$ is a Cauchy sequence in a Banach space, it converges; so let the limit be y. Let T be given by $Tx = [x]$; then $Y_n = Ty_n \to Ty = X$, say. Now we need to show that $Y_n \to X$. Choose $\epsilon > 0$. Then there is a Q such that $p, r > Q \Rightarrow \| X_p - X_r \| < \epsilon/2$ by definition of Cauchy sequence. Then there is an $N > Q$ such that for some n, $N = N_n$ such that $\| X_N - X \| < \epsilon/2$. Further $p > Q \Rightarrow \| X_p - X_N \| < \epsilon/2$. Therefore $p > Q \Rightarrow \| X_p - X \| < \epsilon$.

3. Homomorphisms on Normed Linear Algebras

Definition 3.1. Let L, Q be normed linear algebras; T be a homomorphism from L as a linear space onto Q as a linear space. Then T will be called a *homomorphism* from the algebra L onto the algebra Q if

$$T(XY) = (TX)(TY)$$

Definition 3.2. A *proper* subspace V of a normed linear algebra is called an *ideal* in L if

$$X \in V \quad \text{and} \quad Y \in L \Rightarrow XY \in V \quad \text{and} \quad YX \in V.$$

Note that L itself is not considered an ideal in L.

Theorem 3.1. Let T be a homomorphism from the normed linear algebra L onto the normed linear algebra Q.
Then K_T is an ideal in L.

Proof: Letting $X \in K_T$, $Y \in L$, we must show that $XY \in K_T$ and $YX \in K_T$.
Since $X \in K_T$, $TX = 0$,
so that $T(XY) = TX \cdot TY = 0 \cdot TY = 0$.
Therefore $XY \in K_T$.
Similarly $T(YX) = TY \cdot TX = TY \cdot 0 = 0$
and $YX \in K_T$.

It remains to be shown that $K_T \neq L$. Suppose that $K_T = L$. Then, $e \in K_T$. Hence $Te = 0$ in Q. But Te must be the multiplicative identity in Q. (For $TX \cdot Te = T(Xe) = TX$.) Hence $\| Te \| = 1$, whereas $\| 0 \| = 0$ which is a contradiction.

In what follows let L be a normed linear algebra and let V be an ideal in L.

Theorem 3.2. If $X \equiv X' \mod V$ and $Y \equiv Y' \mod V$ then $XY \equiv X'Y' \mod V$.

Proof: $XY - X'Y' = (XY - XY') + (XY' - X'Y')$

$$= X(Y - Y') + (X - X')Y' \in V$$

The notation $[X]$ will be understood as above.

Definition 3.3. $[X] \cdot [Y] = [XY]$.

Theorem 3.3. If V is an ideal in L, then L/V forms a linear algebra.

Proof: Left to the reader.

Theorem 3.4. The mapping $TX = [X]$ is a homomorphism from L on to L/V with kernel V.

Proof: Left to the reader.

Theorem 3.5. Let S be a homomorphism from L on to Q. Then Q is isomorphic to L/K_S.

Proof: Left to the reader.

Theorem 3.6. Let L be a normed linear algebra and V a closed ideal in L. Then, under the previous definition L/V is a normed linear algebra. Moreover, $\| [X] \| \leq \| X \|$.

Proof: We have only to prove the following statements:

$$\| [e] \| = 1, \qquad \| [X] \| \leq \| X \|$$

and
$$\| [X] \cdot [Y] \| \leq \| [X] \| \cdot \| [Y] \|.$$

But,

$$\| [X] \cdot [Y] \| = \| [XY] \|$$

$$= \inf_{Z \in [XY]} \| Z \|$$

$$\leq \inf_{\substack{Z_1 \in [X] \\ Z_2 \in [Y]}} \| Z_1 Z_2 \|$$

$$\leq \inf_{\substack{Z_1 \in [X] \\ Z_2 \in [Y]}} \| Z_1 \| \cdot \| Z_2 \|$$

$$\leq \inf_{Z_1 \in [X]} \| Z_1 \| \inf_{Z_2 \in [Y]} \| Z_2 \|$$

$$\leq \| [X] \| \cdot \| [Y] \|$$

Also,

$$\| [X] \| = \inf_{Z \in [X]} \| Z \| \leq \| X \|$$

In particular,

$$\| [e] \| \leq \| e \| = 1$$

Moreover $\| [e] \| \neq 0$ since

$$\| [e] \| = 0 \Rightarrow [e] = [0]$$

whereas $[e]$ has an inverse and $[0]$ doesn't.

And,

$$\| [e] \cdot [e] \| \leq \| [e] \| \cdot \| [e] \|$$

so that

$$\| [e] \| \leq \| [e] \|^2$$

i.e.

$$1 \leq \| [e] \|.$$

Since $\quad \| [e] \| \leq 1 \quad$ and $\quad \| [e] \| \geq 1; \quad \| [e] \| = 1.$

Corollary 3.7. If in Theorem 3.6, L is a Banach algebra, so is L/V.

4. Inverses of Elements in Normed Linear Algebras

In this section L is an *Abelian* normed linear algebra.

Theorem 4.1. If $X \in V$ and V is an ideal then X has no inverse.

Proof: Let X have an inverse X^{-1}. Then, $X \in V \Rightarrow e = XX^{-1} \in V$. If $e \in V$, $eY \in V$ for all $Y \in L$ and $V = L$. This contradicts the definition of ideal.

Theorem 4.2. If for every ideal V, $X \notin V$, then X has an inverse.

Proof: Let $Q = \{XY \mid Y \in L\}$ and the following are true:

(1) $X \in Q$,

(2) $XY_1 + XY_2 = X(Y_1 + Y_2)$,

(3) $a(XY) = a(YX) = (aY)X = X(aY)$,

(4) $(XY)Y' = X(YY')$.

(2) and (3) imply that Q is a linear subspace of L.

(2), (3) and (4) imply that Q is an ideal or $Q = L$. By (1) and hypothesis, Q is not an ideal. Hence $Q = L$. In particular, there is a $Y \in L$ such that $XY = e$. Therefore X has an inverse.

Corollary 4.3. X has an inverse $\Leftrightarrow$ X belongs to no ideals.

Proof: Cf. Theorem 4.1 and 4.2.

Definition 4.I. M is a *maximal ideal* in L if:

(1) M is an ideal in L and,

(2) there are no ideals $I \supset M$, $I \neq M$.

Theorem 4.4. For every ideal I, there is a maximal ideal $M \supset I$.

Proof: Let $\mathcal{M}$ be the family defined by:

$$\mathcal{M} = \{J \mid J \supset I \text{ and } J \text{ is an ideal}\}$$

Claim. $\mathcal{M}$ is inductive! Let $\mathcal{C}$ be a chain, $\mathcal{C} \subset \mathcal{M}$. Let $J_0 = \bigcup_{J \in \mathcal{C}} J$. To show that $J_0 \in \mathcal{M}$ we must first prove that J_0 is a superset of I. This can be easily seen from the fact that $J_0 \supset J \supset I$, since J_0 is the union of all sets, $J \in \mathcal{C}$. Next, we must prove that J_0 is an ideal. Let $X, Y \in J_0$. Since J_0 is the union of all sets, $J \in \mathcal{C}$, then $X \in J_X \in \mathcal{C}$ and $Y \in J_Y \in \mathcal{C}$. Since $\mathcal{C}$ is a chain, then either $J_X \subset J_Y$ or $J_Y \subset J_X$. Hence, say, $X, Y \in J_Y$. Hence $X + Y \in J_Y$, since J_Y is an ideal. Hence, $X + Y \in J_0$, since J_0 is the union of all these elements. Next we must show that J_0 is closed under scalar multiplication:

$$X \in J_0$$
$$\Rightarrow X \in J_X$$
$$\Rightarrow aX \in J_X$$
$$\Rightarrow aX \in J_0$$

Hence J_0 is a linear subspace of L. To show that J_0 is an ideal, we must show that J_0 is closed under multiplication from the outside. Let $X \in J_0$ and $Y \in L$. $X \in J_X \in \mathscr{C}$. Since J_X is an ideal,

$$XY = YX \in J_X \in \mathscr{C}$$

Therefore, $XY = YX \in J_0$, since J_0 is the union of all $J \in \mathscr{C}$.

Finally we must show that $J_0 \neq L$. We will do this by contradiction. Suppose $J_0 = L$. Then, in particular, $e \in J_0$. Then $e \in J \in \mathscr{C}$. But e has an inverse, namely, e. Hence, $e \notin J$, because J is an ideal. Therefore this leads to a contradiction. So we have proved the claim that $\mathscr{M}$ is inductive. By Zorn's lemma, $\mathscr{M}$ has a maximal element M. To complete the proof of the theorem, we must prove that M is a maximal ideal. Since $M \in \mathscr{M}$, $M \supset I$ and M is an ideal. Now, wet have only to show that there are no ideals $K \supset M$, $K \neq M$. We will do this by contraction. Suppose K is an ideal and $K \supset M$. Since $M \supset I$, then $K \supset I$. By the definition of $\mathscr{M}$, $K \in \mathscr{M}$. Therefore, $K = M$. Thus we have shown by contradiction that there are no ideals, $K \supset M$, $K \neq M$.

Corollary 4.5. X has an inverse $\Leftrightarrow$ X belongs to no maximal ideal.

Proof $\Rightarrow$: Suppose X has an inverse. This implies that X belongs to no ideal, which in turn implies that X belongs to no maximal ideal.

Proof $\Leftarrow$: Suppose X belongs to no maximal ideal, then X belongs to no ideal (Theorem 4.4). Then X has an inverse.

Theorem 4.6. If I is an ideal, so is $\bar{I}$.

Proof: To prove the theorem we must first prove that $\bar{I}$ is a linear subspace. Let $X, Y \in \bar{I}$. Then, $X_n \to X$ and $Y_n \to Y$ where X_n, $Y_n \in I$; then $X_n + Y_n \in I$. Letting $n \to \infty$, $X + Y \in \bar{I}$. Similarly, if $X \in \bar{I}$:

$X_n \to X$, where $X_n \in I$. Hence, $a \in F$ and $Y \in L \Rightarrow aX_n \in I$ and $X_n Y \in I$.

Letting $n \to \infty$, $aX \in \bar{I}$ and $XY \in \bar{I}$.

Finally, we must prove that $\bar{I} \neq L$. Again, we will prove this by contradiction. Suppose $\bar{I} = L$. Then $e \in \bar{I}$. There exist $\{X_n\}$ with the properties $X_n \in I$, $X_n \to e$. There exists an N such that $\| X_N - e \| < 1$. Therefore, by Theorem 2-17, X_N has an inverse. From Corollary 4.3, $X_N \notin I$. Contradiction.

Corollary 4.7. A maximal ideal is closed.

Proof: $M \subset \overline{M}$. Since M is a maximal ideal, $M = \overline{M}$. Therefore, M is closed.

Definition 4.2. An Abelain linear algebra is a *field*, if $X \neq 0$ implies X has an inverse.

Theorem 4.8. If I is a maximal ideal, then L/I is a field.

Proof (by contradiction): To prove that L/I is a field, we will show that L/I has no ideals except $\{[0]\}$. This will imply that each non-zero element of L/I has an inverse and therefore that L/I is a field. Given I is a maximal ideal. Suppose k is an ideal in L/I, and $k \neq \{[0]\}$. We define:

$$K = \{X \in L \mid [X] \in k\}$$

First, we must show that K is an ideal. To accomplish this, we will show that K satisfies all the closure properties of an ideal. Suppose X_1, $X_2 \in K$. Then $[X_1]$, $[X_2] \in k$. Since k is an ideal, then $[X_1]+[X_2] \in k$. I.e. $[X_1 + X_2] \in k$. Thus, $X_1 + X_2 \in K$.

Next, suppose that $\qquad X \in K$

then $\qquad\qquad\qquad\qquad [X] \in k$

$\qquad\qquad\qquad\qquad\qquad a[X] \in k$

$\qquad\qquad\qquad\qquad\qquad [aX] \in k;$

therefore, $\qquad\qquad\qquad aX \in K.$

Finally, suppose $\qquad\quad X \in K, Y \in L$

then $\qquad\qquad\qquad\qquad [X] \in k$

$\qquad\qquad\qquad\qquad\qquad [X][Y] \in k$

$\qquad\qquad\qquad\qquad\qquad [XY] \in k;$

therefore, $\qquad\qquad\qquad XY \in K.$

Thus, we have shown that all the closure properties of an ideal are valid for K.

Next, we ask, is $K = L$? If it is, then, $e \in K$ which implies $[e] \in k$. Then, $X \in L \Rightarrow [X] = [X][e] \in k$, i.e., $k = L/I$. This is a contradiction, because k is an ideal. Hence, $K \neq L$. We, therefore, conclude that K is an ideal of L.

Now, $Z_0 \in I \Rightarrow [Z_0] = [0] \in k \Rightarrow Z_0 \in K$. I.e, $K \supset I$. But, this is impossible since I is a maximal ideal. This contradiction proves the theorem.

Theorem 4.9. If L/I is a field, then I is maximal.

Proof: Suppose I is not maximal. Then $I \subset K$ and $I \neq K$, where K is an ideal, $K \neq L$.

Let $k = \{[X] \mid X \in K\}$

$[X], [Y] \in k \Rightarrow X, Y \in K \Rightarrow X + Y \in K \Rightarrow [X+Y] \in k \Rightarrow [X] + [Y] \in k$

$[X] \in k \Rightarrow X \in K \Rightarrow aX, XY \in K \Rightarrow [aX], [XY] \in k \Rightarrow a[X],$
$$[X] \cdot [Y] \in k$$

Thus either k is an ideal or $k = L/I$. But since L/I is a field, any ideal is $\{[0]\}$. So either $k = \{[0]\}$ or $k = L/I$.

But $k \neq \{[0]\}$ by the hypothesis $I \neq K$, and $k \neq L/I$ by the hypothesis $K \neq L$. We have arrived at a contradiction, proving that it is impossible that I not be maximal.

Definition 4.3. X is called *regular* if it has an inverse; otherwise it is called *singular*.

Theorem 4.10. In a Banach algebra, the set of all regular elements is an open set.

Proof: Let X be regular.
Let Y satisfy the relation

$$\|X - Y\| < \frac{1}{\|X^{-1}\|}$$

Consider $\| e - X^{-1}Y \| = \| X^{-1}X - X^{-1}Y \| = \| X^{-1}(X - Y) \|$
$$\leq \| X^{-1} \| \cdot \| X - Y \| < 1$$

Therefore, by Theorem 2-1.7, $X^{-1}Y$ has an inverse, say Z

$$X^{-1}YZ = e = X^{-1}ZY$$

But then $X^{-1}Z$ is an inverse of Y, and Y is regular.

Corollary 4.11. The set of singular elements of a Banach algebra is closed. (This set is the union of all maximal ideals.)

Theorem 4.12. Let L be a Banach algebra. Let V be the set of regular elements of L. Let $f(X) = X^{-1}$ for $X \in V$. Then, f is continuous on V.

Lemma. If $X_n \in V$ and $X_n \to e$ then $X_n^{-1} \to e$.

Proof of lemma. There exists an N such that

$$n > N \Rightarrow \| X_n - e \| < \tfrac{1}{2}$$

$$n > N \Rightarrow X_n^{-1} = \sum_{q=0}^{\infty} (e - X_n)^q$$

$$n > N \Rightarrow \| X_n^{-1} \| \leq \sum_{q=0}^{\infty} \| e - X_n \|^q \leq \sum_{q=0}^{\infty} \left(\frac{1}{2}\right)^q = 2$$

Let
$$M = \max(\| X_1^{-1} \|, \| X_2^{-1} \|, \dots, \| X_n^{-1} \|, 2)$$

Then
$$\| X_n^{-1} \| \leq M \text{ for all } n$$

$$\| X_n^{-1} - e \| = \| X_n^{-1} - X_n^{-1} X_n \| = \| X_n^{-1}(e - X_n) \|$$

$$\leq \| X_n^{-1} \| \cdot \| e - X_n \| \leq M \cdot \| e - X_n \| \to 0$$

Proof of Theorem 4.12: Let $X_n \to Z$, $X_n \in V$, $Z \in V$.

Then
$$X_n Z^{-1} \to Z Z^{-1} = e.$$

By the lemma
$$(X_n Z^{-1})^{-1} \to e$$

$$X_n^{-1} Z \to e$$

$$X_n^{-1} Z Z^{-1} \to Z^{-1}$$

$$X_n^{-1} \to Z^{-1}.$$

Problem

Let $L = R^3$, $Q = R$. Let T be defined by: $T(a, b, c) = a$.

(a) Show that T is a homomorphism from L on to Q.

(b) Find K_T.

CHAPTER 5

Analytic Functions into a Banach Space

I. Derivatives

In this chapter L is some fixed Banach space over C, and we shall consider functions α from $D \subset C$ into L.

Definition I.I.
$$\alpha'(z_0) = \lim_{z \to z_0} \frac{1}{z - z_0} [\alpha(z) - \alpha(z_0)]$$
if this limit exists, i.e., if for every $\epsilon > 0$ there is a $\delta > 0$ such that

$$|z - z_0| < \delta \Rightarrow \left\| \alpha'(z_0) - \frac{1}{z - z_0} [\alpha(z) - \alpha(z_0)] \right\| < \epsilon$$

Definition I.2. α is *differentiable* on the set D if $\alpha'(z)$ exists for all $z \in D$.

Theorem I.I. Let α be differentiable on D. Let f be a bounded linear functional on L. Let q be defined on D by $q(z) = f(\alpha(z))$. Then q is differentiable on D and $q'(z) = f(\alpha'(z))$ in D. (Note that q is an ordinary complex-valued function of a complex variable.)

Proof. Let $r_0 = f(\alpha'(z_0))$ for some $z_0 \in D$; consider

$$\left| \frac{q(z) - q(z_0)}{z - z_0} - r_0 \right| = \left| \frac{1}{z - z_0} [f(\alpha(z)) - f(\alpha(z_0))] - f(\alpha'(z_0)) \right|$$

$$= \left| f\left(\frac{1}{z - z_0} [\alpha(z) - \alpha(z_0)] \right) - f(\alpha'(z_0)) \right|$$

$$= \left| f\left(\frac{1}{z - z_0} [\alpha(z) - \alpha(z_0)] - \alpha'(z_0) \right) \right|$$

$$\leq \|f\| \cdot \left\| \frac{1}{z - z_0} [\alpha(z) - \alpha(z_0)] - \alpha'(z_0) \right\| \to 0$$

as $z \to z_0$, by hypothesis.

Theorem 1.2. If α is differentiable in D, then α is continuous in D.

Proof: Let $x_m \in D$; $x_m \to x \in D$.

Let
$$\beta(z) = \frac{1}{z-x}[\alpha(z)-\alpha(x)]-\alpha'(x)$$

(where $z \in D$; $z \neq x$; x and z are complex numbers), and let $\beta(x) = 0$.
Then, $\alpha(z) = \alpha'(x) \cdot (z-x)+\alpha(x)+\beta(z) \cdot (z-x)$. (Note that this equation remains valid for $z = x$.)

$$\alpha(x_m) = \alpha'(x)(x_m-x)+\alpha(x)+\beta(x_m)(x_m-x) \cdot \alpha(x_m) \to$$
$$\alpha'(x) \cdot 0 + \alpha(x)+0 \cdot 0 = \alpha(x), \quad \text{i.e. } \alpha(x_m) \to \alpha(x).$$

2. Integrals of Banach-Space Valued Functions

Let α be a function into L defined on $[a, b]$. We will show how to define

$$\int_a^b \alpha(t)\, dt$$

Definition 2.1. A *partition*, of $[a, b]$ is a finite set of points,
$$P = \{t_1, t_2 \ldots t_{m-1}\}$$
where we take $t_1 < t_2 < t_3 < \ldots < t_{m-1}$; $a < t_i < b$. We usually write $t_0 = a$, $t_m = b$. We write

$$\| P \| = \max_{1 \leq i \leq m} (t_j - t_{i-1})$$

Let α be a function from $[a, b]$ into L.

Definition 2.2. A *Riemann sum* for α over P is a sum:

$$\sum_{i=1}^m \alpha(q_i)(t_i - t_{i-1}) \quad \text{where} \quad t_{i-1} \leq q_i \leq t_i$$

Definition 2.3. α is *Riemann integrable* over $[a, b]$ with *integral I* if the conditions:

(1) $\| P_m \| \to 0$,

(2) for each m, S_m is a Riemann sum for α over P_m imply: $S_m \to I$.

In this case, we write,

$$I = \int_b^a \alpha(t)\, dt$$

We wish to show that if α is continuous on $[a, b]$, then it is Riemann integrable over $[a, b]$.

Lemma I. Let α be continuous on a, b. Then for every $\epsilon > 0$ there is a $\delta > 0$ such that if S, S' are Riemann sums for α over partitions P, P' respectively, and if $\| P \| < \delta$, $\| P' \| < \delta$, then $\| S - S' \| < \epsilon$.

Proof: Let $P'' = P \cup P'$.

Let
$$P = \{t_1, \ t_2, \ t_3, \ldots, t_{n-1}\}$$

$$P' = \{t_1', \ t_2', \ t_3', \ldots, t_{m-1}'\}$$

$$P'' = \{t_1'', \ t_2'', \ t_3'', \ldots, t_{r-1}''\}$$

$$S = \sum_{i=1}^{n} \alpha(q_i)(t_i - t_{i-1}); \quad t_{i-1} \le q_i \le t_i.$$

Similarly
$$S' = \sum_{i=1}^{m} \alpha(q_i')(t_i' - t_{i-1}').$$

Now
$$S = \sum_{k=1}^{r} \alpha(p_k)(t_k'' - t_{k-1}'')$$

where each p_k is one of the q_i.

and
$$S' = \sum_{k=1}^{r} \alpha(p_k')(t_k'' - t_{k-1}'')$$

where each p_k' is one of the q_i'.

Then
$$\| S - S' \| = \left\| \sum_{k=1}^{r} [\alpha(p_k) - \alpha(p_k')](t_k'' - t_{k-1}'') \right\|$$

$$\le \sum_{k=1}^{r} \| \alpha(p_k) - \alpha(p_k') \| (t_k'' - t_{k-1}'')$$

Since $[a, b]$ is a compact set, α is uniformly continuous on $[a, b]$. (Theorem 3-1.6.) Let $\epsilon > 0$ be given. Then there is a number $\eta > 0$ such that, if $t, t' \in [a, b]$ and $|t - t'| < \eta$ then

$$\| \alpha(t) - \alpha(t') \| < \frac{\epsilon}{b-a}$$

Let $\delta = \eta/4$, let $\| P \| < \delta$, $\| P' \| < \delta$. Let $p_k = q_i$. Let $p'_k = q'_j$ where $t_{i-1} \leq p_k \leq t_i$ and $t'_{j-1} \leq p'_k \leq t'_j$. $[t_{i-1}, t_i]$ and $[t'_{j-1}, t'_j]$ overlap, since if they were disjoint, p_k would have some $p'_i \neq p'_k$ corresponding to p_k. Then:

$$|p_k - p'_k| \leq 2\delta < 4\delta = \eta$$

$$\| \alpha(p_k) - \alpha(p'_k) \| < \epsilon/(b-a)$$

$$\| S - S' \| < \sum_{k=1}^{r} \frac{\epsilon}{b-a} (t''_k - t''_{k-1}) = \frac{\epsilon}{b-a} \sum_{k=1}^{r} (t''_k - t''_{k-1})$$

$$= \frac{\epsilon}{(b-a)} (b-a) = \epsilon.$$

Therefore $\| S - S' \| < \epsilon$.

Lemma 2. Let α be continuous on $[a, b]$ and let $\{S_m\}$ be as in Definition 2.2. Then $\{S_m\}$ is a Cauchy sequence.

Proof. Choose $\epsilon > 0$; obtain δ as in Lemma 1. Then there exists N such that $m > N \Rightarrow \| P_m \| < \delta$, since $\| P_m \| \to 0$. Then

$$m, n > N \Rightarrow \| S_m - S_n \| < \epsilon$$

Lemma 3. Let α be continuous on $[a, b]$ let $\{S_n\}$, $\{S'_n\}$ be as in Definition 2.3. Then $S_n - S'_n \to 0$.

Proof: Choose $\epsilon > 0$. Obtain δ as in Lemma 1. Then there exist N_1, N_2 such that

$$n > N_1 \Rightarrow \| P_n \| < \delta; \quad n > N_2 \Rightarrow \| P'_n \| < \delta.$$

Then: $\quad n > \max(N_1, N_2) \Rightarrow \| S_n - S'_n \| < \epsilon.$

Theorem 2.1. If α is continuous on $[a, b]$, α is Riemann integrable on $[a, b]$.

Proof: Immediate from Lemmas 2 and 3.

Theorem 2.2. Let f be a bounded linear functional on L, and let α be a continuous function from $[a, b]$ into L. Then,

$$f\left[\int_a^b \alpha(t)\,dt\right] = \int_a^b f(\alpha(t))\,dt$$

Example: Let L be $C^R(c, d)$

For $X \in L$. let

$$f(X) = \int_c^d X(t)\,dt$$

Let α be a continuous function, $\alpha : [a, b] \to C^R(c, d)$. Setting $\alpha(t) = X_t$, and letting

$$Y(t, u) = X_t(u)$$

it is easily seen that Y is simply a continuous function from

$$[a, b] \times [c, d]$$

into C. Theorem 2.2 asserts about such a Y, that

$$\int_c^d \int_a^b Y(t, u)\,dt\,du = \int_a^b \int_c^d Y(t, u)\,du\,dt$$

i.e. that the order of integrations in an iterated integral may be interchanged.

Proof of Theorem 2.2. Let $t_n \to t$. Then $\alpha(t_n) \to \alpha(t)$, and $f(\alpha(t_n)) \to f(\alpha(t))$.

Therefore $f(\alpha(t))$ is continuous. Let S_n be a sequence of Riemann sums such that

$$S_n \to \int_a^b \alpha(t)\,dt$$

$$S_n = \sum_{i=1}^m \alpha(q_i)(t_i - t_{i-1})$$

where m, q_i, t_i all depend on n. Then

$$f(S_n) = \sum_{i=1}^m f(\alpha(q_i))(t_i - t_{i-1}).$$

Therefore

$$f(S_n) \to \int_a^b f(\alpha(t))\,dt.$$

But,

$$f(S_n) \to f\left(\int_a^b \alpha(t)\,dt\right).$$

3. Line Integrals and Cauchy's Theorem

Definition 3.1. A *smooth Jordan arc* is a function $z(t)$ from $[a, b]$ into C such that

(1) $z'(t)$ is continuous on $[a, b]$

(2) $z(t_2) = z(t_1) \Rightarrow t_2 = t_1$.

Example:

$z(t) = e^{it}$, $0 \leq t \leq \frac{1}{2}\pi$, is a quadrant of a circle.

Definition 3.2. A set $D \subset C$ is called a *Cauchy domain* if:

(1) D is an open set

(2) $z_1, z_2 \in D \Rightarrow$ there exists a smooth Jordan arc which contains z_1 and z_2 and lies entirely in D.

(3) The boundary of D consists of a finite number of smooth Jordan arcs.

Definition 3.3. Let M be a smooth Jordan arc given by $z(t)$, $a \leq t \leq b$. Let α be a continuous function from the points on M into L.

Then,

$$\int_M \alpha(z)\, dz = \int_a^b \alpha(z(t))z'(t)\, dt$$

Definition 3.4. Let α be defined on a set $P \subset C$. Then α is *regular* or *analytic* in P if there is an open set $D \supset P$ on which α is differentiable.

Definition 3.5. Let a curve M be made up of the finite number of smooth Jordan arcs $M_1, M_2, \ldots, M_n$.

Then,

$$\int_M \alpha(z)\, dz = \int_{M_1} \alpha(z)\, dt + \int_{M_2} \alpha(z)\, dz + \ldots + \int_{M_n} \alpha(z)\, dz$$

Theorem 3.1. Let f be a bounded linear functional on L, and let α be continuous on each of the smooth Jordan arcs which make up the curve M. Then

$$f\left(\int_M \alpha(z)\,dz\right) = \int_M f(\alpha(z))\,dz$$

Proof: For M a smooth Jordan arc, the result follows at once from Theorem 2.2 and Definition 3.3. For the general case, we need only employ Definition 3.5 and the linearity of f.

Theorem 3.2. (Classical Cauchy Theorem.) Let α be a complex-valued function which is analytic in the Cauchy domain D plus its boundary M. Then,

$$\int_M \alpha(z)\,dz = 0$$

Proof: See any book on complex variable theory.

Note. A boundary curve is always oriented so that the region is on the left as one advances along the curve.

Theorem 3.3. (*Cauchy's Theorem for Banach Spaces.*) Let α be a Banach-space valued function which is analytic in the Cauchy domain D plus its boundary M. Then,

$$\int_M \alpha(z)\,dz = 0$$

Proof: Let $X_0 = \int_M \alpha(z)\,dz$

Let f be any bounded linear functional on L.

Then
$$f(X_0) = f\left(\int_M \alpha(z)\,dz\right)$$
$$= \int_M f(\alpha(z)\,dz = 0$$

by Theorems 3.2 and 1.1. Hence, by Corollary 3-6.5 (a consequence of the Hahn-Banach Theorem), $X_0 = 0$.

Theorem 3.4. (*Cauchy's integral formula*).   Let α be analytic in the Cauchy domain D plus its boundary M, and let $z \in D$. Then,

$$\alpha(z) = \frac{1}{2\pi i}\int_M \frac{\alpha(w)}{w-z}\,dw$$

Proof: Let

$$X_0 = \alpha(z) - \frac{1}{2\pi i}\int_M \frac{\alpha(w)}{w-z}\,dw$$

Let f be any bounded linear functional on L. Then,

$$f(X_0) = f(\alpha(z)) - \frac{1}{2\pi i}\int_M \frac{f(\alpha(w))}{w-z}\,dw$$
$$= 0$$

by the classical Cauchy integral formula.
Therefore by Corollary 3-6.5, $X_0 = 0$ and

$$\alpha(z) = \frac{1}{2\pi i}\int_M \frac{\alpha(w)}{w-z}\,dw.$$

Lemma. If $\|\,\alpha(z)\,\| \leq M$ on the curve N and the length of N is l then

$$\left\|\int_N \alpha(z)\,dz\right\| \leq Ml$$

Proof: $z = z(t) = x(t) + iy(t)$

$$I = \int_N \alpha(z)\,dz = \int_a^b \alpha[z(t)]z'(t)\,dt.$$

Now $S_n \to I$ where

$$S_n = \sum_{i=1}^m \alpha[z(q_i)]\,z'(q_i)(t_i - t_{i-1}).$$

Also, $$l = \int_a^b \sqrt{[x'(t)]^2 + [y'(t)]^2}\,dt = \int_a^b |z'(t)|\,dt.$$

Hence $$\|S_n\| \leq \sum_{i=1}^m \|\alpha[z(q_i)]\| \cdot |z'(q_i)| \cdot (t_i - t_{i-1})$$

$$\leq M \sum_{i=1}^m |z'(q_i)|(t_i - t_{i-1}).$$

And letting $n \to \infty$,

$$\|I\| \leq M \int_a^b |z'(t)|\,dt = Ml.$$

Theorem 3.5. (*Liouville's Theorem.*) If α is analytic in the entire complex plane and $\| \alpha(z) \| \leq M$, then α is constant.

Proof 1 (*Using the classical Liouville's Theorem*): Suppose α is not constant; then there exist z_1 and z_2 such that $\alpha(z_1) \neq \alpha(z_2)$. Hence there must be a bounded linear functional f on L such that

$$f(\alpha(z_1)) \neq f(\alpha(z_2))$$

by Corollary 3-6.6. But $f(\alpha(z))$ is analytic in the complex plane, by Theorem 1.1.

Now $|f(\alpha(z))| \leq \|f\| \cdot \|\alpha(z)\| \leq M \|f\|$

Hence by the classical Liouville's Theorem, $f(\alpha(z))$ is constant. This is a contradiction; hence α must be constant.

Proof 2 (*Using Cauchy's integral formula*): Suppose α is not constant; then there exist z_1 and z_2 such that $\alpha(z_1) \neq \alpha(z_2)$. Let N be a circle with center at z_1 and radius r so large that $|z_2 - z_1| < r$. Then by Theorem 3.4

$$\alpha(z_2) = \frac{1}{2\pi i} \int_N \frac{\alpha(w)}{w - z_2} dw$$

$$\alpha(z_1) = \frac{1}{2\pi i} \int_N \frac{\alpha(w)}{w - z_1} dw$$

Hence, $\quad \|\alpha(z_2) - \alpha(z_1)\| = \frac{1}{2\pi} \left\| \int_N \left[\frac{1}{w - z_2} - \frac{1}{w - z_1} \right] \alpha(w)\, dw \right\|$

$$= \frac{1}{2\pi} \left\| \int_N \frac{(z_2 - z_1)\alpha(w)}{(w - z_2)(w - z_1)} dw \right\|$$

Now, $\qquad \|\alpha(w)\| \leq M, \quad |w - z_1| = r$, and

$|w - z_2| = |(w - z_1) - (z_2 - z_1)| \geq ||w - z_1| - |z_2 - z_1|| = r - |z_2 - z_1|$

Hence by the Lemma

$$\|\alpha(z_2) - \alpha(z_1)\| \leq \frac{1}{2\pi} \frac{|z_2 - z_1| M \cdot 2\pi r}{[r - |z_2 - z_1|]r} \to 0 \quad \text{as} \quad r \to \infty$$

Hence $\alpha(z_2) = \alpha(z_1)$ which is a contradiction and α must therefore be constant.

4. Banach Algebras which are Fields

To say that a Banach algebra L is a *field* is to say that L is Abelian and that for $X \in L$, $X \neq O$, X has an inverse, or what comes to the same thing:

If X, $U \in L$, $X \neq O$, $\exists V \in L$, $V = U/X$, i.e. $VX = XV = U$.

Theorem 4.1. A Banach algebra over C, which is also a field, is isomorphic to the field of complex numbers C.

Proof: Common hypotheses for all lemmas: L is a Banach algebra over C, which is also a field.

Lemma I. For each $X \in L$, there exists an $a \in C$ such that $X = ae$.

Proof: Suppose that there exists an $X \in L$ such that $X \neq ae$ for any $a \in C$. Let $\alpha(z) = (X - ze)^{-1}$. Note that $X - ze$ is non-zero by hypothesis and the inverse exists because L is a field.

$$\alpha'(z_0) = \lim_{z \to z_0} \frac{(X - ze)^{-1} - (X - z_0 e)^{-1}}{z - z_0} = \lim_{z \to z_0} \frac{(z - z_0)e}{(z - z_0)(X - ze)(X - z_0 e)}$$

$$= [(X - z_0 e)^{-1}]^2$$

Also, $\qquad (X - ze)^{-1} = z^{-1}(Xz^{-1} - e)^{-1} \quad$ for $\quad z \neq 0$.

Now, as $|z| \to \infty$, $Xz^{-1} \to 0$, i.e., $Xz^{-1} - e \to -e$.

By Theorem 4–4.12 (continuity of the inverse function)

$$(Xz^{-1} - e)^{-1} \to (-e)^{-1} = -e$$

Hence $\qquad z^{-1}(Xz^{-1} - e)^{-1} \to 0 \cdot (-e) = 0 \quad$ as $\quad |z| \to \infty$

Therefore, there is a number $p > 0$ such that

$$|z| > p \Rightarrow \| (X - ze)^{-1} \| < 1$$

But $\{z \mid \; |z| \leq p\}$ is compact. Hence by Corollary 3–1.2 there is a number K such that $|z| \leq p \Rightarrow \|(X-ze)^{-1}\| \leq K$. But this means that $\alpha(z)$ is analytic everywhere and $\|\alpha(z)\| \leq \max(K, 1)$.

Therefore α is constant by Liouville's Theorem.

Note that $\alpha(z) \to 0$ as $z \to \infty$.

Hence $\alpha(z) = 0$. But this means that $(X-ze)^{-1} = 0$ and this implies that $e = (X-ze)(X-ze)^{-1} = 0$. This is impossible; hence the lemma is proved.

Lemma 2. The mapping $a \leftrightarrow ae$ is an isomorphism between C and L.

Proof: Let $a \leftrightarrow ae$, $b \leftrightarrow be$, then $a+b \leftrightarrow (a+b)e = ae+be$,

$$a \cdot b \leftrightarrow (a \cdot b)e = (ae) \cdot (be) = a(be)$$

$$|a| = \|ae\| \quad \text{since} \quad \|e\| = 1;$$

finally suppose $ae = be$ and $a \neq b$.

This implies $(a-b)e = 0$, $(a-b) \neq 0$, hence $e = 0$ which is impossible. Therefore $ae = be \Rightarrow a = b$ and $a \leftrightarrow ae$ is an isomorphism between C and L.

Theorem 4.1 follows at once from Lemma 2.

Corollary 4.2. Let L be a Banach algebra over C and let M be a maximal ideal in L. Then L/M is isomorphic to C.

Proof: L/M is a Banach algebra which is a field, (Theorem 4-4.6). Hence L/M is isomorphic to C.

5. The Convolution Algebra l^1

Definition 5.1.

$$l^1 = \left\{ \{a_n\} \mid n = 0, \pm 1, \pm 2, \ldots \text{ and } a_n \in C \text{ and } \sum_{n=-\infty}^{\infty} |a_n| < \infty \right\}$$

An example of such an $\{a_n\}$ is the sequence where

$$a_n = \frac{1}{n^2+1}:$$

$$\ldots, \frac{1}{10}, \frac{1}{5}, \frac{1}{2}, 1, \frac{1}{2}, \frac{1}{5}, \frac{1}{10}, \ldots$$

In this space we introduce the definitions:

(1) $\{a_n\} + \{b_n\} = \{a_n + b_n\}$

(2) $\quad c\{a_n\} = \{ca_n\}$

(3) $\quad \|\{a_n\}\| = \sum_{n=-\infty}^{\infty} |a_n|$

(4) $\{a_n\}\{b_n\} = \{c_n\}$, where $c_n = \sum_{p=-\infty}^{\infty} a_p b_{n-p}$

Theorem 5.1. Under the operations just defined, l^1 is a Banach algebra.

Proof: The proof that l^1 is a linear algebra is trivial. Notice that

$$0 = \{a_n\} \text{ where } a_n = 0 \text{ for all } n$$

$$e = \{\delta_n^0\} \text{ where } \delta_n^m = \begin{cases} 1 \text{ for } n = m \\ 0 \text{ for } n \neq m \end{cases}$$

For, $\qquad\qquad \{a_n\}\{\delta_n^0\} = \{c_n\}$

where $\qquad\qquad c_n = \sum_{p=-\infty}^{\infty} a_n \delta_{n-p} = a_n \cdot 1 = a_n.$

$$\begin{aligned}
\|\{a_n\} + \{b_n\}\| &= \|\{a_n + b_n\}\| \\
&= \sum |a_n + b_n| \\
&\leq \sum [|a_n| + |b_n|] \\
&= \sum |a_n| + \sum |b_n| \\
&= \|\{a_n\}\| + \|\{b_n\}\|
\end{aligned}$$

where we have omitted the index and limits of summation.

$$\| \{a_n\} \cdot \{b_n\} \| = \| \{c_n\} \|$$
$$= \sum_n \left| \sum_p a_p \, b_{n-p} \right|$$
$$\leq \sum_n \sum_p |a_p| |b_{n-p}|$$
$$= \sum_p \sum_n |a_p| |b_{n-p}|$$
$$= \sum_p \left[|a_p| \sum_n |b_{n-p}| \right]$$
$$= \sum_p \left[|a_p| \sum_n |b_n| \right]$$
$$= \sum_p |a_p| \cdot \sum_n |b_n|$$
$$= \| \{a_n\} \| \cdot \| \{b_n\} \|$$
$$\| e \| = \sum_n |\delta_n^0| = 1$$

Thus l^1 is a normed linear algebra. Now we show it is also complete.

Let $\{X_n\}$ be a sequence, $X_n \in l^1$, and suppose

$$\lim_{\substack{n \to \infty \\ p \to \infty}} \| X_n - X_p \| = 0$$

Now let each $X_n = \{a_m^{(n)}\}$. Then

$$\lim_{\substack{n \to \infty \\ p \to \infty}} \sum_m \left| a_m^{(n)} - a_m^{(p)} \right| = 0$$

For fixed m,

$$\left| a_m^{(n)} - a_m^{(p)} \right| \leq \sum_m \left| a_m^{(n)} - a_m^{(p)} \right| \to 0$$

Thus

$$\lim_{\substack{n \to \infty \\ p \to \infty}} \left| a_m^{(n)} - a_m^{(p)} \right| = 0$$

uniformly in m.

It then follows that there must exist a sequence $X = \{b_m\}$ such that $\lim_{n \to \infty} a_m^{(n)} = b_m$ uniformly in m.

$$\| X_n - X \| = \sum_n \left| a_m^{(n)} - b_m \right|$$

Therefore,

$$\lim_{n \to \infty} \| X_n - X \| = \lim_{n \to \infty} \lim_{q \to \infty} \sum_{m=-q}^{q} \left| a_m^{(n)} - b_m \right|$$

$$= \lim_{q \to \infty} \lim_{n \to \infty} \sum_{m=-q}^{q} \left| a_m^{(n)} - b_m \right|$$

$$= \lim_{q \to \infty} 0 = 0$$

where the interchange of limit operations is justified by uniformity.

Also,

$$\lim_{n \to \infty} \left| a_m^{(n)} \right| = \left| b_m \right| \text{ uniformly in } m.$$

$$\sum_m \left| b_m \right| = \sum_m \lim_{n \to \infty} \left| a_m^{(n)} \right|$$

$$= \lim_{n \to \infty} \sum_m \left| a_m^{(n)} \right|$$

$$= \lim_{n \to \infty} \| \{a_n\} \| < \infty$$

which proves the completeness of l^1 and completes the proof of the Theorem.

Let us introduce $G = \{\delta_n^1\} \in l^1$. We assert the following identity:

$$\{a_n\} = \sum_{n=-\infty}^{\infty} a_n G^n$$

Let us verify this identity. Certainly

$$G^{-1} = \{\delta_n^{-1}\}$$

for

$$\{\delta_n^1\} \cdot \{\delta_n^{-1}\} = \{c_n\}$$

where

$$c_n = \sum_p \delta_p^1 \delta_{n-p}^{-1} = 0$$

except when $p = 1$ and $n - p = -1$, and thus when $n = 0$. Therefore $c_n = \delta_n^0$, and $\{c_n\} = \{\delta_n^0\} = e$. Also,

$$G^2 = \{c_n\}, \quad c_n = \sum_p \delta_p^1 \delta_{n-p}^1 = \delta_n^2 \text{ and } G^k = \{\delta_n^k\}$$

Now let us compute

$$\sum_{k=-r}^{r} a_k G^k$$

$$= \{\ldots, 0, 0, a_{-r}, a_{-r+1}, \ldots, a_0, a_1, a_2, \ldots, a_{r-1}, a_r, 0, 0, \ldots\}$$

Thus, letting $r \to \infty$, we obtain the desired result.

Let T be any homomorphism from l^1 on to C. Let $TG = t$, a complex number. Now $\| G \| = 1$, and remembering that

$$\| [X] \| = \inf_{z \in [X]} \| z \| \leq \| X \|$$

we have $| t | \leq 1$.

$TG^{-1} = 1/t$, and since $\| G^{-1} \| = 1$, it also follows that $| 1/t | \leq 1$. Thus $| t | = 1$.

We may represent $TG = t = e^{i\theta}$. Then

$$T\{a_n\} = T \sum_{n=-\infty}^{\infty} a_n G^n$$

$$= T \lim_{r \to \infty} \sum_{n=-r}^{r} a_n G^n$$

$$= \lim_{r \to \infty} T \sum_{n=-r}^{r} a_n G^n$$

$$= \lim_{r \to \infty} \sum_{n=-r}^{r} a_n (TG)^n$$

$$= \sum_{r=-\infty}^{\infty} a_n e^{ni\theta}$$

Which elements $\{a_n\} \in l^1$ have inverses? Those $\{a_n\}$ which belong to no maximal ideals (cf. Corollary 4–4.5). In other words, those $\{a_n\}$ which belong to the kernel of no homomorphisms from l^1 onto C, i.e., those $\{a_n\}$ which are mapped into 0 by no homomorphism.

Conclusion: $\{a_n\}$ has an inverse $\Leftrightarrow$ there is no θ, $0 \leq \theta < 2\pi$, such that

$$\sum_n a_n e^{ni\theta} = 0$$

This leads to:

Theorem 5.2. (Wiener.) Let

$$f(t) = \sum_{n=-\infty}^{\infty} a_n e^{nit}, \quad 0 \leq t < 2\pi$$

where

$$\sum_{n=-\infty}^{\infty} | a_n | < \infty$$

Let $f(t)$ vanish for *no* t, $0 \leqq t < 2\pi$. Then there are numbers b_n such that

$$\frac{1}{f(t)} = \sum_{n=-\infty}^{\infty} b_n e^{nit}, \quad \sum_{n=-\infty}^{\infty} |b_n| < \infty, \quad 0 \leqq t < 2\pi.$$

Proof: Let $X = \{a_n\}$, $X \in l^1$. By the conclusion above, X has an inverse in l^1,

$$X^{-1} = \{b_n\}, \quad \sum_{n=-\infty}^{\infty} |b_n| < \infty$$

Let

$$g(t) = \sum_n b_n e^{int}$$

$$g(t) f(t) = \sum_{n=-\infty}^{\infty} a_n e^{int} \cdot \sum_{n=-\infty}^{\infty} b_n e^{int}$$

$$= \sum_{n=-\infty}^{\infty} \sum_{p=-\infty}^{\infty} a_p b_{n-p} e^{int}$$

$$= \sum_{n=-\infty}^{\infty} \delta_n^0 e^{int} = 1.$$

Let us define a sub-algebra of l^1:

Definition 5.2.

$$A = \{\{a_n\} \,|\, \{a_n\} \in l^1 \quad \text{and} \quad a_n = 0 \quad \text{for} \quad n < 0\}$$

We have easily

Corollary 5.3. A is a closed sub-algebra of l^1 and hence is also a Banach algebra.

In the algebra A,

$$\{a_n\} = \sum_{n=0}^{\infty} a_n G^n$$

What does a homomorphism T from A on to C look like? Let

$$TG = z \in C$$

$$|z| \leqq \|G\| = 1$$

$$T\{a_n\} = \sum_{n=0}^{\infty} a_n z^n, \quad |z| \leqq 1$$

Thus as in the proof of Theorem 5.2, we have:

Theorem 5.4. Let

$$f(z) = \sum_{n=0}^{\infty} a_n z^n \quad \text{for} \quad |z| \leq 1$$

where

$$\sum_{n=0}^{\infty} |a_n| < \infty$$

and let $f(z)$ have no zeros for $|z| \leq 1$. Then

$$\frac{1}{f(z)} = \sum_{n=0}^{\infty} b_n z^n, \quad \sum_{n=0}^{\infty} |b_n| < \infty$$

Example.

$$f(z) = 2 + \sum_{n=1}^{\infty} \frac{z^n}{n^2}$$

Problems

1. Complete the proof of Theorem 5.4,

2. (*a*) Show that under the definition:

$$(X \cdot Y)(t) = X(t) \cdot Y(t)$$

the space $C^C(0, 1)$ (which we already know is a Banach space) becomes a Banach algebra.

    (*b*) Show that for each maximal ideal M in $C^C(0, 1)$, there is a point t_0 in $[0, 1]$, such that

$$M = \{X \mid X(t_0) = 0\}$$

(*Hint:* Use the Weierstrass approximation theorem which states: Let X be continuous on $[a, b]$. Then, there is a sequence $\{P_n\}$ of *polynomials* such that $P_n(t) \to X(t)$ *uniformly* on $[a, b]$.)

Index

A CATALOG OF SELECTED

DOVER BOOKS
IN SCIENCE AND MATHEMATICS

Mathematics–Bestsellers

HANDBOOK OF MATHEMATICAL FUNCTIONS: with Formulas, Graphs, and Mathematical Tables, Edited by Milton Abramowitz and Irene A. Stegun. A classic resource for working with special functions, standard trig, and exponential logarithmic definitions and extensions, it features 29 sets of tables, some to as high as 20 places. 1046pp. 8 x 10 1/2. 0-486-61272-4

ABSTRACT AND CONCRETE CATEGORIES: The Joy of Cats, Jiri Adamek, Horst Herrlich, and George E. Strecker. This up-to-date introductory treatment employs category theory to explore the theory of structures. Its unique approach stresses concrete categories and presents a systematic view of factorization structures. Numerous examples. 1990 edition, updated 2004. 528pp. 6 1/8 x 9 1/4. 0-486-46934-4

MATHEMATICS: Its Content, Methods and Meaning, A. D. Aleksandrov, A. N. Kolmogorov, and M. A. Lavrent'ev. Major survey offers comprehensive, coherent discussions of analytic geometry, algebra, differential equations, calculus of variations, functions of a complex variable, prime numbers, linear and non-Euclidean geometry, topology, functional analysis, more. 1963 edition. 1120pp. 5 3/8 x 8 1/2. 0-486-40916-3

INTRODUCTION TO VECTORS AND TENSORS: Second Edition–Two Volumes Bound as One, Ray M. Bowen and C.-C. Wang. Convenient single-volume compilation of two texts offers both introduction and in-depth survey. Geared toward engineering and science students rather than mathematicians, it focuses on physics and engineering applications. 1976 edition. 560pp. 6 1/2 x 9 1/4. 0-486-46914-X

AN INTRODUCTION TO ORTHOGONAL POLYNOMIALS, Theodore S. Chihara. Concise introduction covers general elementary theory, including the representation theorem and distribution functions, continued fractions and chain sequences, the recurrence formula, special functions, and some specific systems. 1978 edition. 272pp. 5 3/8 x 8 1/2. 0-486-47929-3

ADVANCED MATHEMATICS FOR ENGINEERS AND SCIENTISTS, Paul DuChateau. This primary text and supplemental reference focuses on linear algebra, calculus, and ordinary differential equations. Additional topics include partial differential equations and approximation methods. Includes solved problems. 1992 edition. 400pp. 7 1/2 x 9 1/4. 0-486-47930-7

PARTIAL DIFFERENTIAL EQUATIONS FOR SCIENTISTS AND ENGINEERS, Stanley J. Farlow. Practical text shows how to formulate and solve partial differential equations. Coverage of diffusion-type problems, hyperbolic-type problems, elliptic-type problems, numerical and approximate methods. Solution guide available upon request. 1982 edition. 414pp. 6 1/8 x 9 1/4. 0-486-67620-X

VARIATIONAL PRINCIPLES AND FREE-BOUNDARY PROBLEMS, Avner Friedman. Advanced graduate-level text examines variational methods in partial differential equations and illustrates their applications to free-boundary problems. Features detailed statements of standard theory of elliptic and parabolic operators. 1982 edition. 720pp. 6 1/8 x 9 1/4. 0-486-47853-X

LINEAR ANALYSIS AND REPRESENTATION THEORY, Steven A. Gaal. Unified treatment covers topics from the theory of operators and operator algebras on Hilbert spaces; integration and representation theory for topological groups; and the theory of Lie algebras, Lie groups, and transform groups. 1973 edition. 704pp. 6 1/8 x 9 1/4. 0-486-47851-3

CATALOG OF DOVER BOOKS

A SURVEY OF INDUSTRIAL MATHEMATICS, Charles R. MacCluer. Students learn how to solve problems they'll encounter in their professional lives with this concise single-volume treatment. It employs MATLAB and other strategies to explore typical industrial problems. 2000 edition. 384pp. 5 3/8 x 8 1/2. 0-486-47702-9

NUMBER SYSTEMS AND THE FOUNDATIONS OF ANALYSIS, Elliott Mendelson. Geared toward undergraduate and beginning graduate students, this study explores natural numbers, integers, rational numbers, real numbers, and complex numbers. Numerous exercises and appendixes supplement the text. 1973 edition. 368pp. 5 3/8 x 8 1/2. 0-486-45792-3

A FIRST LOOK AT NUMERICAL FUNCTIONAL ANALYSIS, W. W. Sawyer. Text by renowned educator shows how problems in numerical analysis lead to concepts of functional analysis. Topics include Banach and Hilbert spaces, contraction mappings, convergence, differentiation and integration, and Euclidean space. 1978 edition. 208pp. 5 3/8 x 8 1/2. 0-486-47882-3

FRACTALS, CHAOS, POWER LAWS: Minutes from an Infinite Paradise, Manfred Schroeder. A fascinating exploration of the connections between chaos theory, physics, biology, and mathematics, this book abounds in award-winning computer graphics, optical illusions, and games that clarify memorable insights into self-similarity. 1992 edition. 448pp. 6 1/8 x 9 1/4. 0-486-47204-3

SET THEORY AND THE CONTINUUM PROBLEM, Raymond M. Smullyan and Melvin Fitting. A lucid, elegant, and complete survey of set theory, this three-part treatment explores axiomatic set theory, the consistency of the continuum hypothesis, and forcing and independence results. 1996 edition. 336pp. 6 x 9. 0-486-47484-4

DYNAMICAL SYSTEMS, Shlomo Sternberg. A pioneer in the field of dynamical systems discusses one-dimensional dynamics, differential equations, random walks, iterated function systems, symbolic dynamics, and Markov chains. Supplementary materials include PowerPoint slides and MATLAB exercises. 2010 edition. 272pp. 6 1/8 x 9 1/4. 0-486-47705-3

ORDINARY DIFFERENTIAL EQUATIONS, Morris Tenenbaum and Harry Pollard. Skillfully organized introductory text examines origin of differential equations, then defines basic terms and outlines general solution of a differential equation. Explores integrating factors; dilution and accretion problems; Laplace Transforms; Newton's Interpolation Formulas, more. 818pp. 5 3/8 x 8 1/2. 0-486-64940-7

MATROID THEORY, D. J. A. Welsh. Text by a noted expert describes standard examples and investigation results, using elementary proofs to develop basic matroid properties before advancing to a more sophisticated treatment. Includes numerous exercises. 1976 edition. 448pp. 5 3/8 x 8 1/2. 0-486-47439-9

THE CONCEPT OF A RIEMANN SURFACE, Hermann Weyl. This classic on the general history of functions combines function theory and geometry, forming the basis of the modern approach to analysis, geometry, and topology. 1955 edition. 208pp. 5 3/8 x 8 1/2. 0-486-47004-0

THE LAPLACE TRANSFORM, David Vernon Widder. This volume focuses on the Laplace and Stieltjes transforms, offering a highly theoretical treatment. Topics include fundamental formulas, the moment problem, monotonic functions, and Tauberian theorems. 1941 edition. 416pp. 5 3/8 x 8 1/2. 0-486-47755-X

Browse over 9,000 books at www.doverpublications.com

Mathematics–Logic and Problem Solving

PERPLEXING PUZZLES AND TANTALIZING TEASERS, Martin Gardner. Ninety-three riddles, mazes, illusions, tricky questions, word and picture puzzles, and other challenges offer hours of entertainment for youngsters. Filled with rib-tickling drawings. Solutions. 224pp. 5 3/8 x 8 1/2. 0-486-25637-5

MY BEST MATHEMATICAL AND LOGIC PUZZLES, Martin Gardner. The noted expert selects 70 of his favorite "short" puzzles. Includes The Returning Explorer, The Mutilated Chessboard, Scrambled Box Tops, and dozens more. Complete solutions included. 96pp. 5 3/8 x 8 1/2. 0-486-28152-3

THE LADY OR THE TIGER?: and Other Logic Puzzles, Raymond M. Smullyan. Created by a renowned puzzle master, these whimsically themed challenges involve paradoxes about probability, time, and change; metapuzzles; and self-referentiality. Nineteen chapters advance in difficulty from relatively simple to highly complex. 1982 edition. 240pp. 5 3/8 x 8 1/2. 0-486-47027-X

SATAN, CANTOR AND INFINITY: Mind-Boggling Puzzles, Raymond M. Smullyan. A renowned mathematician tells stories of knights and knaves in an entertaining look at the logical precepts behind infinity, probability, time, and change. Requires a strong background in mathematics. Complete solutions. 288pp. 5 3/8 x 8 1/2.

0-486-47036-9

THE RED BOOK OF MATHEMATICAL PROBLEMS, Kenneth S. Williams and Kenneth Hardy. Handy compilation of 100 practice problems, hints and solutions indispensable for students preparing for the William Lowell Putnam and other mathematical competitions. Preface to the First Edition. Sources. 1988 edition. 192pp. 5 3/8 x 8 1/2. 0-486-69415-1

KING ARTHUR IN SEARCH OF HIS DOG AND OTHER CURIOUS PUZZLES, Raymond M. Smullyan. This fanciful, original collection for readers of all ages features arithmetic puzzles, logic problems related to crime detection, and logic and arithmetic puzzles involving King Arthur and his Dogs of the Round Table. 160pp. 5 3/8 x 8 1/2. 0-486-47435-6

UNDECIDABLE THEORIES: Studies in Logic and the Foundation of Mathematics, Alfred Tarski in collaboration with Andrzej Mostowski and Raphael M. Robinson. This well-known book by the famed logician consists of three treatises: "A General Method in Proofs of Undecidability," "Undecidability and Essential Undecidability in Mathematics," and "Undecidability of the Elementary Theory of Groups." 1953 edition. 112pp. 5 3/8 x 8 1/2. 0-486-47703-7

LOGIC FOR MATHEMATICIANS, J. Barkley Rosser. Examination of essential topics and theorems assumes no background in logic. "Undoubtedly a major addition to the literature of mathematical logic." – *Bulletin of the American Mathematical Society*. 1978 edition. 592pp. 6 1/8 x 9 1/4. 0-486-46898-4

INTRODUCTION TO PROOF IN ABSTRACT MATHEMATICS, Andrew Wohlgemuth. This undergraduate text teaches students what constitutes an acceptable proof, and it develops their ability to do proofs of routine problems as well as those requiring creative insights. 1990 edition. 384pp. 6 1/2 x 9 1/4. 0-486-47854-8

FIRST COURSE IN MATHEMATICAL LOGIC, Patrick Suppes and Shirley Hill. Rigorous introduction is simple enough in presentation and context for wide range of students. Symbolizing sentences; logical inference; truth and validity; truth tables; terms, predicates, universal quantifiers; universal specification and laws of identity; more. 288pp. 5 3/8 x 8 1/2. 0-486-42259-3

Mathematics–Algebra and Calculus

VECTOR CALCULUS, Peter Baxandall and Hans Liebeck. This introductory text offers a rigorous, comprehensive treatment. Classical theorems of vector calculus are amply illustrated with figures, worked examples, physical applications, and exercises with hints and answers. 1986 edition. 560pp. 5 3/8 x 8 1/2.	0-486-46620-5

ADVANCED CALCULUS: An Introduction to Classical Analysis, Louis Brand. A course in analysis that focuses on the functions of a real variable, this text introduces the basic concepts in their simplest setting and illustrates its teachings with numerous examples, theorems, and proofs. 1955 edition. 592pp. 5 3/8 x 8 1/2.	0-486-44548-8

ADVANCED CALCULUS, Avner Friedman. Intended for students who have already completed a one-year course in elementary calculus, this two-part treatment advances from functions of one variable to those of several variables. Solutions. 1971 edition. 432pp. 5 3/8 x 8 1/2.	0-486-45795-8

METHODS OF MATHEMATICS APPLIED TO CALCULUS, PROBABILITY, AND STATISTICS, Richard W. Hamming. This 4-part treatment begins with algebra and analytic geometry and proceeds to an exploration of the calculus of algebraic functions and transcendental functions and applications. 1985 edition. Includes 310 figures and 18 tables. 880pp. 6 1/2 x 9 1/4.	0-486-43945-3

BASIC ALGEBRA I: Second Edition, Nathan Jacobson. A classic text and standard reference for a generation, this volume covers all undergraduate algebra topics, including groups, rings, modules, Galois theory, polynomials, linear algebra, and associative algebra. 1985 edition. 528pp. 6 1/8 x 9 1/4.	0-486-47189-6

BASIC ALGEBRA II: Second Edition, Nathan Jacobson. This classic text and standard reference comprises all subjects of a first-year graduate-level course, including in-depth coverage of groups and polynomials and extensive use of categories and functors. 1989 edition. 704pp. 6 1/8 x 9 1/4.	0-486-47187-X

CALCULUS: An Intuitive and Physical Approach (Second Edition), Morris Kline. Application-oriented introduction relates the subject as closely as possible to science with explorations of the derivative; differentiation and integration of the powers of x; theorems on differentiation, antidifferentiation; the chain rule; trigonometric functions; more. Examples. 1967 edition. 960pp. 6 1/2 x 9 1/4.	0-486-40453-6

ABSTRACT ALGEBRA AND SOLUTION BY RADICALS, John E. Maxfield and Margaret W. Maxfield. Accessible advanced undergraduate-level text starts with groups, rings, fields, and polynomials and advances to Galois theory, radicals and roots of unity, and solution by radicals. Numerous examples, illustrations, exercises, appendixes. 1971 edition. 224pp. 6 1/8 x 9 1/4.	0-486-47723-1

AN INTRODUCTION TO THE THEORY OF LINEAR SPACES, Georgi E. Shilov. Translated by Richard A. Silverman. Introductory treatment offers a clear exposition of algebra, geometry, and analysis as parts of an integrated whole rather than separate subjects. Numerous examples illustrate many different fields, and problems include hints or answers. 1961 edition. 320pp. 5 3/8 x 8 1/2.	0-486-63070-6

LINEAR ALGEBRA, Georgi E. Shilov. Covers determinants, linear spaces, systems of linear equations, linear functions of a vector argument, coordinate transformations, the canonical form of the matrix of a linear operator, bilinear and quadratic forms, and more. 387pp. 5 3/8 x 8 1/2.	0-486-63518-X

Mathematics–History

THE WORKS OF ARCHIMEDES, Archimedes. Translated by Sir Thomas Heath. Complete works of ancient geometer feature such topics as the famous problems of the ratio of the areas of a cylinder and an inscribed sphere; the properties of conoids, spheroids, and spirals; more. 326pp. 5 3/8 x 8 1/2.   0-486-42084-1

THE HISTORICAL ROOTS OF ELEMENTARY MATHEMATICS, Lucas N. H. Bunt, Phillip S. Jones, and Jack D. Bedient. Exciting, hands-on approach to understanding fundamental underpinnings of modern arithmetic, algebra, geometry and number systems examines their origins in early Egyptian, Babylonian, and Greek sources. 336pp. 5 3/8 x 8 1/2.   0-486-25563-8

THE THIRTEEN BOOKS OF EUCLID'S ELEMENTS, Euclid. Contains complete English text of all 13 books of the Elements plus critical apparatus analyzing each definition, postulate, and proposition in great detail. Covers textual and linguistic matters; mathematical analyses of Euclid's ideas; classical, medieval, Renaissance and modern commentators; refutations, supports, extrapolations, reinterpretations and historical notes. 995 figures. Total of 1,425pp. All books 5 3/8 x 8 1/2.

Vol. I: 443pp.   0-486-60088-2
Vol. II: 464pp.   0-486-60089-0
Vol. III: 546pp.   0-486-60090-4

A HISTORY OF GREEK MATHEMATICS, Sir Thomas Heath. This authoritative two-volume set that covers the essentials of mathematics and features every landmark innovation and every important figure, including Euclid, Apollonius, and others. 5 3/8 x 8 1/2.   Vol. I: 461pp.   0-486-24073-8
Vol. II: 597pp.   0-486-24074-6

A MANUAL OF GREEK MATHEMATICS, Sir Thomas L. Heath. This concise but thorough history encompasses the enduring contributions of the ancient Greek mathematicians whose works form the basis of most modern mathematics. Discusses Pythagorean arithmetic, Plato, Euclid, more. 1931 edition. 576pp. 5 3/8 x 8 1/2.
0-486-43231-9

CHINESE MATHEMATICS IN THE THIRTEENTH CENTURY, Ulrich Libbrecht. An exploration of the 13th-century mathematician Ch'in, this fascinating book combines what is known of the mathematician's life with a history of his only extant work, the Shu-shu chiu-chang. 1973 edition. 592pp. 5 3/8 x 8 1/2.
0-486-44619-0

PHILOSOPHY OF MATHEMATICS AND DEDUCTIVE STRUCTURE IN EUCLID'S ELEMENTS, Ian Mueller. This text provides an understanding of the classical Greek conception of mathematics as expressed in Euclid's Elements. It focuses on philosophical, foundational, and logical questions and features helpful appendixes. 400pp. 6 1/2 x 9 1/4.   0-486-45300-6

BEYOND GEOMETRY: Classic Papers from Riemann to Einstein, Edited with an Introduction and Notes by Peter Pesic. This is the only English-language collection of these 8 accessible essays. They trace seminal ideas about the foundations of geometry that led to Einstein's general theory of relativity. 224pp. 6 1/8 x 9 1/4. 0-486-45350-2

HISTORY OF MATHEMATICS, David E. Smith. Two-volume history – from Egyptian papyri and medieval maps to modern graphs and diagrams. Non-technical chronological survey with thousands of biographical notes, critical evaluations, and contemporary opinions on over 1,100 mathematicians. 5 3/8 x 8 1/2.
Vol. I: 618pp.   0-486-20429-4
Vol. II: 736pp.   0-486-20430-8